Christian Wasmer

Atomic-Resolution Structures of Amyloid Fibrils by Solid-State NMR

Christian Wasmer

Atomic-Resolution Structures of Amyloid Fibrils by Solid-State NMR

Südwestdeutscher Verlag für Hochschulschriften

Impressum/Imprint (nur für Deutschland/only for Germany)
Bibliografische Information der Deutschen Nationalbibliothek: Die Deutsche Nationalbibliothek verzeichnet diese Publikation in der Deutschen Nationalbibliografie; detaillierte bibliografische Daten sind im Internet über http://dnb.d-nb.de abrufbar.
Alle in diesem Buch genannten Marken und Produktnamen unterliegen warenzeichen-, marken- oder patentrechtlichem Schutz bzw. sind Warenzeichen oder eingetragene Warenzeichen der jeweiligen Inhaber. Die Wiedergabe von Marken, Produktnamen, Gebrauchsnamen, Handelsnamen, Warenbezeichnungen u.s.w. in diesem Werk berechtigt auch ohne besondere Kennzeichnung nicht zu der Annahme, dass solche Namen im Sinne der Warenzeichen- und Markenschutzgesetzgebung als frei zu betrachten wären und daher von jedermann benutzt werden dürften.

Verlag: Südwestdeutscher Verlag für Hochschulschriften GmbH & Co. KG
Heinrich-Böcking-Str. 6-8, 66121 Saarbrücken, Deutschland
Telefon +49 681 37 20 271-1, Telefax +49 681 37 20 271-0
Email: info@svh-verlag.de

Approved by: Zürich, ETH, Diss., 2011

Herstellung in Deutschland:
Schaltungsdienst Lange o.H.G., Berlin
Books on Demand GmbH, Norderstedt
Reha GmbH, Saarbrücken
Amazon Distribution GmbH, Leipzig
ISBN: 978-3-8381-3026-2

Imprint (only for USA, GB)
Bibliographic information published by the Deutsche Nationalbibliothek: The Deutsche Nationalbibliothek lists this publication in the Deutsche Nationalbibliografie; detailed bibliographic data are available in the Internet at http://dnb.d-nb.de.
Any brand names and product names mentioned in this book are subject to trademark, brand or patent protection and are trademarks or registered trademarks of their respective holders. The use of brand names, product names, common names, trade names, product descriptions etc. even without a particular marking in this works is in no way to be construed to mean that such names may be regarded as unrestricted in respect of trademark and brand protection legislation and could thus be used by anyone.

Publisher: Südwestdeutscher Verlag für Hochschulschriften GmbH & Co. KG
Heinrich-Böcking-Str. 6-8, 66121 Saarbrücken, Germany
Phone +49 681 37 20 271-1, Fax +49 681 37 20 271-0
Email: info@svh-verlag.de

Printed in the U.S.A.
Printed in the U.K. by (see last page)
ISBN: 978-3-8381-3026-2

Copyright © 2011 by the author and Südwestdeutscher Verlag für Hochschulschriften GmbH & Co. KG and licensors
All rights reserved. Saarbrücken 2011

dedicated to my parents

meinen Eltern

"It is far better to grasp the universe as it really is than to persist in delusion, however satisfying and reassuring."

Carl Edward Sagan (1934–1996)

Contents

Abstract 11

Zusammenfassung 13

Introduction 15
    1   Solid-State NMR 16
    2   Biological Systems in Solid-State NMR 18
    3   Prions and Amyloid Fibrils 20
    4   The HET-s Prion 20

I The Structure of HET-s(218-289) 23
    1   Introduction 24
    2   Materials and Methods 25
        2.1   Preparation of isotope-labeled HET-s(218-289) fibrils. ... 25
        2.2   Solid-state NMR experiments. 25
        2.3   Structure calculation. 26
    3   Spectra and Distance Restraints 36
    4   The Structure 38
    5   Back-calculations 41

II Full-length HET-s 45
    1   Introduction 46
    2   Methods 48
        2.1   Sample preparation 48
        2.2   Solid-State NMR 49
    3   Results 49
    4   Discussion 59

III Non-Infectious pH 3 Fibrils — 63

1. Introduction — 64
2. Materials and Methods — 65
   - 2.1 Sample Preparation — 65
   - 2.2 Experimental details — 65
3. Solid-State NMR on pH 3 Fibrils — 65
   - 3.1 CP-based experiments — 65
   - 3.2 INEPT-based experiments — 68

IV HET-s(218-289) Inclusion Bodies — 71

1. Introduction — 72
2. Materials and Methods — 73
   - 2.1 Inclusion body purification — 73
   - 2.2 Sample preparation for EM — 73
   - 2.3 Samples for solid-state NMR and H/D-exchange — 74
   - 2.4 Infectivity assays — 75
   - 2.5 Solid-state NMR spectroscopy — 75
   - 2.6 H/D-exchange by liquid-state NMR — 75
3. Bio-Chemical Experiments — 76
4. Solid-State NMR — 77
5. Conclusions — 79

V A HET-s Homologue—FgHET-s — 83

1. Introduction — 84
2. Materials and Methods — 85
   - 2.1 Plasmids and strains — 85
   - 2.2 Protein expression — 85
   - 2.3 Protein purification — 85
   - 2.4 ThT-binding determination — 86
   - 2.5 Electron microscopy — 86
   - 2.6 Aggregation assays — 87
   - 2.7 Chemical denaturation curves — 87
   - 2.8 Hydrogen-Deuterium Exchange — 88
   - 2.9 Solid-state NMR — 88
3. Results — 90
   - 3.1 Homologues of HET-s — 90
   - 3.2 FgHET-s(218-289) forms amyloid fibrils — 90

		3.3	FgHET-s(218-289) fibrils seed HET-s(218-289)	93
		3.4	H/D-exchange	93
		3.5	Solid-state NMR	95
		3.6	Cross-seeded fibrils adopt a similar structure as unseeded.	101
	4	Discussion and Conclusions		101
		4.1	Structural comparison to HET-s(218-289).	101
		4.2	Evolutionary conservation of the β-solenoid fold.	105
		4.3	Structural similarity accounts for cross-seeding	106
	5	Summary		106

VI The Structure of FgHET-s(218-289)   109

	1	Introduction		110
	2	Materials and Methods		111
		2.1	Sample Preparation	111
		2.2	Solid-state NMR	112
		2.3	Implementation of the Structure Calculation	113
	3	Results		114
		3.1	Spectra and Distance Restraints	114
		3.2	Complementary Restraints—H/D exchange and TALOS .	116
		3.3	The Structure Calculation	116
		3.4	The Structure of FgHET-s(218-289)	122
	4	Conclusions and Discussion		124
		4.1	The Structure Calculation	124
		4.2	The Structure	125

Conclusions and Outlook   129

X Appendix   131

	1	Non-infectious pH 3 Fibrils		132
	2	Full-length HET-s		138
	3	HET-s(218-289) Inclusion Bodies		142
	4	FgHET-s(218-289)		148
	5	Sample preparation protocol		163
		5.1	Recombinant Expression	164
		5.2	Purification	166
		5.3	Fibrillation	169
		5.4	Buffers and Media	172

Bibliography	**174**
Acknowledgments	**185**
Curriculum Vitae	**187**
Publications	**191**

Abstract

Prions are infectious proteins capable of self-replicating their conformation and are best known as the agent of a number of diseases including scrapie in sheep, bovine spongiform encephalopathy (BSE) in cattle, and new variant Creutzfeldt-Jakob disease in humans. Prions have also been found and described in yeast and filamentous fungi. The infectious form of most prions has been identified or is at least being conjectured to be a β-sheet-rich molecular aggregate termed amyloid fibril. Additionally, amyloid formation is eponymous for a group of diseases termed Amyloidoses that includes several highly prevalent neurodegenerative diseases such as Alzheimer's and Parkinson's disease. Due to their decisive role in diseases, it is of great interest to understand the stability and formation of amyloid fibrils in detail. To achieve this, it is in turn required to determine the atomic-resolution molecular structures of such entities. However, structure determination of protein fibrils is mostly hampered by the fact that they are neither soluble nor crystallizable and therefore not amenable to the two most successful techniques of experimental structural biology—solution nuclear magnetic resonance (NMR) and X-ray crystallography. Accordingly, prior to this study no structure of a prion in the infectious amyloid state had been determined. Solid-state NMR is not limited to crystals or solutes, as even amorphous, insoluble samples may be investigated and it is therefore the method of choice for the structural characterization of amyloid fibrils. However, determination of protein structures by solid-state NMR is still a laborious process and has only been achieved for few, mostly crystalline, proteins so far.

For the determination of the structure of HET-s(218-289), the amyloid core of a prion of the filamentous fungus *Podospora anserina*, a large number of NMR samples with different isotopic labeling schemes were produced. These enabled the recording of NMR spectra that selectively reflect either the inter- or intramolecular order within the fibrils, which proved to be crucial information for the structure determination. Based on the experimental restraints, we found that HET-s(218-289) forms a left-handed β-solenoid, with each molecule forming two helical windings. It exhibits a compact hydrophobic core, at least 23 hydrogen bonds, three salt bridges, and two asparagine ladders per molecule. The obtained structure gives important insights on the features leading to the extraordinarily high stability of these amyloid fibrils.

Abstract

The structure of the prion domain in isolation was put into the context of the full-length HET-s protein, which is the naturally occurring entity. To this purpose, a comparison of the solid-state NMR spectra of fibrils of HET-s and the prion domain (218-289) was performed. The collected data show that the C-terminal residues 218-289 within HET-s adopt the same β-solenoid fold as in isolation, or, in other words, the HET-s(218-289) amyloid fold is preserved in full-length HET-s fibrils. From the further analysis of the NMR spectra, a molecular model of the HET-s fibrils could be developed. Therein, the N-terminal domain adopts a poorly ordered molten-globule state, while the C-terminal part is highly organized into the amyloid fibril core observed for HET-s(218-289).

The prion domain HET-s(218-289) has recently been shown to form amyloid fibrils at a pH below 3.5 *in vitro*, which, in contrast to those formed at neutral pH, show little or no prion infectivity. This enabled us to address, on a molecular level, the differences that distinguish infectious from noninfectious polymorphs of the same polypeptide. The solid-state NMR spectra recorded on HET-s(218-289) pH 3 fibrils show that their structure differs extensively from that of the infectious pH 7 form. The structural dissimilarities of the two fibrillar species explain why the pH 3 fibrils cannot induce the prion form of HET-s, i.e. why they are not infectious.

When recombinantly expressed in *E. coli*, HET-s(218-289) readily aggregates into inclusion bodies. This is a well known phenomenon that occurs when expressing insoluble proteins in bacterial cells and is of interest because it may be closely related to the formation of amyloid deposits associated with many neurodegenerative diseases. The structure of inclusion bodies in general had not yet been investigated on an atomic scale and they were widely regarded as disordered protein aggregates. Our collected data unequivocally prove the amyloid nature of HET-s(218-289) inclusion bodies and show them to have the same highly organized structure as the *in vitro* amyloid fibrils. This shows that inclusion bodies can equal amyloid fibrils on an atomic scale.

Finally, we structurally investigated FgHET-s(218-289), a distant homologue of HET-s from the filamentous fungus *Fusarium graminearum* by solid-state NMR. The sequence-specific resonance assignment for almost all observable residues was derived from a set of NMR spectra. The chemical shifts together with the hydrogen/deuterium-exchange rates detected by solution NMR already show that the overall folds of FgHET-s(218-289) and HET-s(218-289) are indeed very similar. By recording and evaluating solid-state NMR spectra containing structural information in the form of distance restraints, we could determine the atomic-resolution structure of the amyloid fibrils formed by FgHET-s(218-289). It confirms a pronounced structural similarity to HET-s(218-289) but also some interesting differences that explain the different chemical properties of FgHET-s(218-289). Additionally, to determine this structure, an improved protocol for the structure calculation of protein fibrils by solid-state NMR was developed that greatly reduces both the required amount of protein and the number of NMR spectra to be recorded.

Zusammenfassung

Prionen, infektiöse Proteine die in der Lage sind ihre Struktur selbst zu replizieren, sind vor allem bekannt als Erreger einiger Krankheiten, darunter die Traberkrankheit bei Schafen, bovine spongiforme Enzephalopathie (BSE) bei Rindern und die 'new variant' der Creutzfeldt-Jakob-Erkrankung bei Menschen. Prionen wurden aber auch schon in Hefe und Fadenpilzen entdeckt. Als ansteckende Form wird bei den meisten Prionen ein β-Faltblatt-reiches Aggregat vermutet, welches Amyloidfibrille gennant wird. Die Ablagerung von Amyloiden ist dabei namengebend für eine ganze Klasse von Krankheiten, den sogenannten Amyloidosen, zu denen auch häufig vorkommende neurodegenerative Erkrankungen wie Alzheimer und Parkinson gehören. Aufgrund ihrer Verbindung zu diesen Krankheiten ist die Bestimmung atomar aufgelöster Strukturen von Amyloidfibrillen von grossem Interesse, da so deren Bildung und Stabilität besser verstanden werden kann. Die Strukturbestimmung von Proteinfibrillen wird jedoch dadurch erschwert, dass diese weder löslich noch kristallisierbar, und damit auch durch die beiden erfolgreichsten Methoden der experimentellen Strukturbiologie—Lösungs-NMR (nukleare magnetische Resonanz) und Röntgenkristallografie—nicht charakterisierbar sind. Dementsprechend war auch bis jetzt noch keine Struktur eines Prions in der infektiösen Amyloidfaltung bekannt. Da Festkörper-NMR auch auf unlösliche, amorphe Proben anwendbar ist, stellt es die Methode der Wahl zur Strukturbestimmung von Amyloidfibrillen dar. Allerdings ist dies zur Zeit noch ein aufwendiger Prozess, und nur wenige, meist kristalline, Proteine wurden bis jetzt auf diese Art strukturell charakterisiert.

Zur Bestimmung der Struktur von HET-s(218-289), dem Amyloidkern eines Prions des Fadenpilzes *Podospora anserina*, wurden zahlreiche NMR-Proben mit verschiedenen Isotopenmarkierungsschemata hergestellt. Mit diesen konnten wir NMR-Spektren aufnehmen, die spezifische Informationen über die intra- und intermolekulare Ordnung der Amyloidfibrillen enthalten, welche sich für die Strukturbestimmung als unentbehrlich erwiesen. Basierend auf den experimentellen Daten konnten wir herausfinden, dass HET-s(218-289) einen sogenannten β-Solenoiden mit zwei Windungen pro Molekül bildet. Dieser weist einen dreieckigen hydrophoben Kern sowie mindestens 23 Wasserstoffbrücken, drei Salzbrücken und zwei Asparaginleitern pro Molekül auf und zeigt damit eindrucksvoll wie die hohe Stabilität der Amyloidfibrillen erreicht wird.

Zusammenfassung

Um die Struktur des natürlich vorkommenden HET-s-Proteins ganzer Länge zu untersuchen, verglichen wir die Festkörper-NMR-Spektren von Fibrillen aus HET-s und jenen aus HET-s(218-289). Diese Daten zeigen, dass die C-terminalen Aminosäurereste 218-289 von HET-s den gleichen β-Solenoiden bilden wie in Isolation und damit die Struktur der Priondomäne in HET-s erhalten ist. Wir konnten ausserdem ein molekulares Modell für die HET-s-Fibrillen ableiten, in welchem sich die N-terminale Domäne in einem schlecht geordneten 'molten-globule'-Zustand befindet, der C-terminale Teil jedoch hochgeordnet ist.

Die Priondomäne (218-289) von HET-s kann, wie kürzlich gezeigt wurde, auch eine andere Art Amyloidfibrillen bilden, die bei pH-Werten von unter 3.5 entsteht und wenig oder keine der für Prionen typischen Infektiosität aufweist. Dadurch konnten wir die Unterschiede zwischen der infektiösen und der nichtinfektiösen Form des selben Polypeptids auf molekularer Ebene untersuchen. Die Festkörper-NMR-Spektren der sogenannten pH 3-Fibrillen zeigen, dass sich diese strukturell erheblich von der infektiösen pH 7-Form unterscheiden. Die Unterschiede zwischen den beiden, vom gleichen Protein gebildeten, Fibrillenarten können erklären, weshalb die pH 3-Fibrillen nicht die Prionenform von HET-s hervorrufen und nicht ansteckend sind.

Bei der rekombinanten Expression von HET-s(218-289) in *E. coli* Bakterien wird das Protein in sogenannten Einschlusskörperchen abgelagert. Dies ist ein wohl bekanntes Phänomen, das bei der rekombinanten Expression unlöslicher Proteine in Bakterien auftritt und mit der Bildung von Amyloidablagerungen im Kontext neurodegenerativer Erkrankungen verwandt sein könnte. Einschlusskörperchen werden meist als amorphe Ablagerungen angesehen, wurden aber bis jetzt noch nicht mit atomarer Auflösung untersucht. Unsere Daten, aufgenommen an Einschlusskörperchen von HET-s(218-289), zeigen eindeutig, dass diese Amyloide sind und die hochgeordnete Struktur der *in vitro* gebildeten Fibrillen besitzen. Wir konnten damit zeigen, dass Einschlusskörperchen Amyloidfibrillen sein können.

FgHET-s(218-289), ein zu HET-s homologes Protein aus dem Fadenpilz *Fusarium graminearum* bildet ebenfalls Fibrillen, die wir mit Festkörper-NMR untersucht haben. Mit einem Set von Spektren konnten wir die Sequenz-spezifische Resonanzzuordnung fast aller beobachtbarer Aminosäurereste erreichen. Zusammen mit den Wasserstoff/Deuterium-Austauschraten zeigen diese Daten bereits, dass sich FgHET-s(218-289) und HET-s(218-289) in der Tat sehr ähnlich sind. Aufgrund interatomarer Abstandsbegrenzungen, die aus weiteren Festkörper-NMR-Spektren extrahiert wurden, bestimmten wir die Struktur von FgHET-s(218-289) mit atomarer Auflösung. Diese ist der von HET-s(218-289) beeindruckend ähnlich, weist aber im Detail einige Unterschiede auf, die auch die unterschiedlichen Eigenschaften dieser Fibrillen erklären können. Im Rahmen dieser Strukturbestimmung wurde zusätzlich ein verbessertes Protokoll zur Berechnung von Amyloidstrukturen basierend auf Festkörper-NMR-Daten entwickelt, welches mit einem stark reduzierten Satz an Proteinproben und NMR-Spektren durchführbar ist.

Introduction

1 Solid-State NMR

Solution NMR is one of the most powerful techniques in structural biology that is able to characterize both the average structure and the dynamics of bio-molecules, e.g. proteins. As a technique that is dependent on fast molecular tumbling in solution, it is however limited to proteins or protein assemblies below a certain size. Several experimental techniques like transverse relaxation-optimized spectroscopy (Pervushin et al. (1997)) and the use of deuterated protein samples (LeMaster and Richards (1988)), have vastly extended the range of solution NMR to bigger and therefore slower-tumbling molecules. Nevertheless, large molecular assemblies in the MDa-range are still inaccessible by this technique, even using the most advanced experimental setups available today.

Solid-state NMR on the contrary does not know this limitation as it does not require molecular tumbling. In general, in NMR of spin-1/2 nuclei the Hamiltonian has the form

$$\hat{\mathcal{H}} = \hat{\mathcal{H}}_Z + \hat{\mathcal{H}}_{CS} + \hat{\mathcal{H}}_J + \hat{\mathcal{H}}_D,$$

where $\hat{\mathcal{H}}_Z$ is the Zeeman Hamiltonian, $\hat{\mathcal{H}}_{CS}$ the full chemical-shift Hamiltonian, $\hat{\mathcal{H}}_J$ the J-coupling Hamiltonian, and $\hat{\mathcal{H}}_D$ the dipolar-coupling Hamiltonian. In solution, the orientation dependent (anisotropic) part of the chemical shift tensor as well as the dipolar couplings would be averaged to zero on the timescale of the molecular tumbling. It is these two interactions that will render the solid-state NMR spectrum of a static powder broad and difficult to interpret, as the crystallites in the powder all assume different orientations. Especially for larger systems, such as proteins, with numerous observable spins this so-called powder pattern renders it impossible to obtain meaningful structural information about the sample under investigation. A solution to overcome this problem has been suggested by Lowe (1959), who rotated the NMR sample around an axis inclined by $\theta_M \approx 54.7°$ against the static magnetic field $\vec{B}_0$. The "magic angle" θ_M fulfills the condition $3\cos^2\theta_M - 1 = 0$, which leads to an averaging to zero of the anisotropic parts of the Hamiltonian in a first-order approximation. In order to obtain a considerable attenuation of a specific term, the magic angle spinning (MAS) frequency has to be larger than the respective anisotropic interaction. This can be achieved for the anisotropic part of the chemical-shift Hamiltonian $\hat{\mathcal{H}}_{CS}$ and the dipolar couplings $\hat{\mathcal{H}}_D$ between low-γ nuclei such as ^{13}C or ^{15}N, but for dense networks of protons with their large dipolar couplings even the highest MAS frequencies achievable today (70 kHz) are insufficient. Therefore high-resolution ^{1}H solid-state NMR is only possible by partly deuterating the sample and thereby randomly removing most ^{1}H–^{1}H dipolar couplings. (Current works on this subject have been done by Linser et al. (2008) and Zhou et al. (2007).)

An additional complication of solid-state NMR arises from the dipolar couplings between low- and high-γ nuclei. Although the dipolar couplings in between low-γ nuclei are efficiently attenuated by MAS, the coupling to a network of strongly coupled high-γ nuclei—usually protons—persists and leads to a substantial broadening of the NMR lines. To overcome this effect, the proton spins can be decoupled by applying an RF-field at their resonance frequency. Today, mostly pulsed decoupling sequences such as TPPM (Bennett et al. (1995)), SPINAL (Fung et al. (2000)) and XiX (Detken et al. (2002)) are used.

As the residual dipolar couplings under MAS between nuclei, especially protons, lead to a fast decay of transverse magnetization, polarization transfer through J-couplings based on the INEPT-scheme (Morris and Freeman (1979); Burum and Ernst (1980))—the preferred method in solution NMR—is rendered impossible or at least difficult. A notable work, where this problem has been overcome, was done by Chen et al. (2007), although the experimental conditions (150 kHz ^{1}H decoupling for > 100 ms) may not be beneficial for the life-span of most probes and protein samples. Due to these problems, the most commonly used heteronuclear transfer technique in solid-state NMR is Hartmann-Hahn cross-polarization (CP) (Hartmann and Hahn (1962); Hediger et al. (1994, 1995); Baldus et al. (1996)) that aims to re-introduce dipolar couplings between specific nuclei by RF-irradiation. It can be used for both polarization transfer from protons to low-γ nuclei and in between two different low-γ nuclei.

However, a CP-step can only be accomplished between different kinds of nuclei, i.e. nuclei with a sufficiently large difference in their resonance frequencies. To transfer polarization between nuclei of a kind, numerous methods are available. TOtal through-Bond correlation SpectroscopY (TOBSY) (Baldus and Meier (1996); Hardy et al. (2001, 2003)) aims at suppressing homo- and heteronuclear dipolar couplings and thereby yields a transfer based mainly on the homonuclear J-couplings. The DREAM scheme (Verel et al. (2001)) transfers polarization via direct homonuclear couplings, and is most commonly employed to transfer polarization in between carbon atoms. These couplings are reintroduced by an adiabatic passage through the HORROR-condition, i.e. by irradiating on the according nucleus with an RF-amplitude of 1/2 of the MAS frequency. The resulting spectra show mainly contacts between strongly coupled nuclei, while transfer between weakly coupled atoms is suppressed by dipolar truncation. Therefore, DREAM-spectra of low-γ nuclei are mainly useful to identify intra-residual connectivities, i.e. spin-systems in the context of the sequence-specific resonance assignment.

The following homonuclear transfer schemes all employ a second kind of nucleus, which is in all cases ^{1}H. The simplest and probably also most popular form of a homonuclear correlation spectrum is achieved by Proton-Driven Spin-Diffusion (PDSD) (Goldman and Jacquinot (1982); Grommek et al. (2006)) or the akin Dipolar-Assisted Rotational Resonance (DARR) (Takegoshi et al. (2001)) and MIxed Rotational and ROtary-Resonance (MIRROR) (Scholz et al. (2009)). The two latter experiments share one pulse sequence (but not one theoretical basis), in which an RF-field is applied at the proton frequency during the mixing period. While these kind of spectra, in principle do not suffer from dipolar truncation, the resulting spectra are still largely dominated by short-range contacts, which hampers the observation of larger numbers of long-range interactions that would in turn be required for determining the structure of a protein.

The following techniques have been proposed to specifically obtain such long-range interactions. The CHHC and NHHC sequences (Lange et al. (2002, 2003)) use an explicit transfer to ^{1}H and back to the low-γ nucleus in question (^{15}N or ^{13}C) by short CP-steps. The mixing itself takes place directly in between protons (proton spin diffusion) that retain sufficiently strongly dipolar couplings even under fast MAS. Another mechanism that has recently been developed is Third-Spin Assisted Recoupling (TSAR). It can be used both homo- and heteronuclearly and is then more specifically called Proton-Assisted Recoupling (PAR) and Proton Assisted Insensitive Nuclei CP (PAIN-CP), respectively (Lewandowski et al. (2007); De Paepe et al. (2006)).

2 Biological Systems in Solid-State NMR

The first structure of a protein, calculated from a set of restraints derived from solid-state NMR spectra was presented for a micro-crystalline preparation of the α-spectrin SH3 domain by Castellani et al. (2002). By expressing the protein in media containing either 2-^{13}C glycerol or 1,3-^{13}C labelled glycerol, an alternating labeling scheme that was initially invented to enable the reliable determination of ^{13}C relaxation data (LeMaster and Kushlan (1996)), was obtained. Herein, two neighboring carbon atoms are rarely ^{13}C labeled, which strongly reduces the peril of relayed transfers that render PDSD spectra difficult to interpret when recorded on uniformly labelled samples. Thereby, the quality of the long-range restraints derived from such spectra is enhanced. Additionally, almost no ^{13}C–^{13}C J-couplings are present, which leads to narrower resonances. On the other hand, the dilution of spin labels may be regarded as a disadvantage, as only about $(1/2)^2 = 1/4$ of the expected cross-peaks are theoretically observable in a 2D spectrum for a single sample. The structure calculation itself was performed with the program CNS (Brünger et al. (1998)) and yielded a root mean square deviation (RMSD) from the mean structure of 1.6 Å for the C$^\alpha$-atoms of the 15 lowest-energy structures out of 200. The RMSD to the previously known X-ray structure was 2.6 Å.

The structure of a peptide fragment of the amyloidogenic protein transthyretin, TTR(105-115) has been calculated from measurements of both distance restraints and backbone torsion angles by Jaroniec et al. (2004). It could be shown, that this peptide adopts an extended β-sheet structure within a fibril that shows a high degree of long-range order.

Kaliotoxin, a 38-residue peptide found in the venom of the scorpion *Androctonus mauretaincus mauretanicus*, binds to and thereby blocks voltage-dependent eukaryotic potassium channels (Grissmer et al. (1994)). Distance restraints were extracted from CHHC and NHHC spectra (Lange et al. (2002, 2003)) to obtain the structure of Kaliotoxin in its free (Lange et al. (2005); Korukottu et al. (2008)) and bound state (Lange et al. (2006))

The helical transmembrane photosynthetic light-harvesting 2 protein complex has been studied also employing the CHHC experiment (Ganapathy et al. (2007)). The authors obtained several distance restraints providing structural information on the active site.

The Y145Stop mutant of the human Prion protein (hPrP) is associated with a hereditary amyloid disease. The resonances observable by CP-based experiments have been sequentially assigned and arise from 29 amino acid residues at the C-terminus (Helmus et al. (2008)). Most of the remaining 116 residues are dynamically disordered as apparent in INEPT-based spectra (Helmus et al. (2010)).

The ubiquitous protein Ubiquitin was employed as a model system and it could be shown that—by a multi-cycle structural refinement process—one can obtain the structure of a small (70 residues) micro-crystalline protein by using solely PDSD spectra and one uniformly labelled sample (Manolikas et al. (2008)).

In very recent years, numerous additional studies were carried out on interesting biological systems that are not listed here.

3 Prions and Amyloid Fibrils

Prions are infectious proteins capable of self-replicating their conformation and are best known as the agent of diseases such as scrapie in sheep (Prusiner (1982)), bovine spongiform encephalopathy in cattle (Wells et al. (1987)), and a new variant of Creutzfeldt-Jakob disease in humans (Will et al. (1996)). Prions have also been described in yeast (Wickner (1994); Wickner et al. (2007)) and filamentous fungi (Benkemoun et al. (2006)) (For a review see (Baxa et al. (2006))). The infectious form of most known prions has been characterized as a β sheet–rich molecular aggregate termed an amyloid fibril (Baxa et al. (2005); Kajava and Steven (2006)). At least for the yeast prion Ure2p, there is contradicting evidence (and believes), if the prion form of this protein is in fact an amyloid fibrils (Shewmaker et al. (2009)) or another kind of fibril (Bousset et al. (2002); Loquet et al. (2009)). In any case, as the sole difference between the infectious prion and the non-prion form is the protein fold, it is of strong interest to determine the structure of both. While the soluble non-prion form can principle be characterized by the two major techniques of structural biology—X-ray crystallography and solution NMR (see e.g. Riek et al. (1996); Hornemann et al. (1997) and Zahn et al. (2000))—protein fibrils are generally not accessible by these techniques.

A few prion protein fragments consisting of less than 7 residues do form crystals and were structurally analyzed by Sawaya et al. (2007). It is however unclear, how these protein crystals relate to the actual amyloid fibrils. As solid-state NMR has no principal limitations regarding molecular weight and long-range order of the sample, it is ideally suited to obtain structural information about amyloid fibrils (Tycko (2006)). It may also be noted at this point that amyloid fibrils are not necessarily linked to disease and may also have a function (Chiti and Dobson (2006); Fowler et al. (2007)) and of course the techniques described in this thesis are also applicable to obtain structural knowledge about these. The review of Greenwald and Riek (2010) summarizes the most recent findings on the structure and function of amyloid fibrils.

4 The HET-s Prion

HET-s is a prion of the filamentous fungus *Podospora anserina* (Coustou et al. (1997); Rizet (1952)). In the non-prion state HET-s behaves as a soluble protein. It consists of a well-ordered, globular N-terminal domain comprising 227 amino-acid residues and a highly dynamic C-terminal part in random-coil conformation

(Balguerie et al. (2003); Ritter et al. (2005)). The structure of the globular domain HET-s(1-227) alone has been determined by X-ray crystallography (Greenwald et al. (2010)) and this part has been shown to be identically structured in the full-length soluble HET-s (Balguerie et al. (2003)). The fold of the globular domain is almost exclusively α-helical and has been termed a 'HeLo'-domain (Greenwald et al. (2010)).

In its aggregated form, HET-s has a proteinase K-resistant core formed by the C-terminal residues 218 to 289. These residues alone have been shown to form infectious amyloid fibrils in vitro and they are also necessary for the prion infectivity of HET-s (Balguerie et al. (2003)). Therefore this part has been termed the prion domain of HET-s. Earlier work showed that HET-s(218-289) in its fibrillar state consists of four β-strands forming two windings of a β-solenoid (Ritter et al. (2005)). However, it does not contain atomic-resolution structural information and no information about the intermolecular β-sheet propagation (parallel or antiparallel). Besides, the remains of about 220 N-terminal residues in the amyloid fold of HET-s remained elusive.

The *het-s/het-S* locus in the filamentous fungus Podospora anserina has two naturally occurring alleles, *het-s* and *het-S*, each encoding for two proteins of 289 amino acid residues and differing in only 13 residues evenly distributed over the sequence (Turcq et al. (1991)). The *het-S* allele encodes for the HET-S protein, which, although it also contains a prion domain, has so far been found exclusively in a soluble non-prion state. Fungal cells bearing the *het-s* allele can display one of two phenotypes: those with the non-prion phenotype [Het-s*] behave neutrally towards cells bearing the *het-S* allele, while those displaying the [Het-s] phenotype, i.e. containing the aggregated prion form of HET-s, display an antagonistic behavior when confronted with a *het-S* strain. The resulting cell-death-like reaction is termed heterokaryon incompatibility. This self/non-self recognition phenomenon has been suspected to prevent or alleviate the effects of different forms of parasitism (Saupe (2007)). It is however not clear, if this can be considered the main and/or only function of HET-s. A more detailed description of this process and the most current findings on it can be found in Greenwald et al. (2010).

Introduction

Chapter I

The Structure of HET-s(218-289)

This work was done in collaboration with Hélène Van Melckebeke and Adam Lange and is published (Wasmer et al. (2008a)). Adam Lange recorded and evaluated the CHHC spectra; Hélène Van Melckebeke implemented and performed the structure calcukations. An extensive analysis of the structure calculation has been performed by Hélène Van Melckebeke (Van Melckebeke et al. (2010)).

1 Introduction

We used solid-state NMR to determine the structure of the rigid core of the fibrils formed by HET-s(218-289). Previous work on amyloid fibrils by NMR has shown that structural information can be obtained from these noncrystalline entities (Tycko (2006); Ferguson et al. (2006); Jaroniec et al. (2004); Heise et al. (2005); Shewmaker et al. (2006); Petkova et al. (2006)). Resonance assignment of HET-s(218-289) solid-state NMR spectra have been described previously (Siemer et al. (2006b); Siemer (2006)). We determined a large number of distance restraints from uniformly labeled samples and specifically identify purely intra- and intermolecular restraints by using differently labeled samples. The ^{1}H-^{1}H and ^{13}C-^{13}C internuclear distance restraints were derived from ^{13}C-detected proton-spin diffusion [carbon-proton-proton-carbon (CHHC) and nitrogen-proton-proton-carbon (NHHC) experiments (Lange et al. (2002, 2005))] and from proton-driven ^{13}C spin diffusion (PDSD) (Szeverenyi et al. (1982)), respectively (for a list of experiments, see Table I.1). Part of a CHHC spectrum from a uniformly isotopically labeled (U-[^{13}C, ^{15}N]) sample is shown in Fig. I.4A as a representative example. From this spectrum, 41 structurally meaningful restraints were obtained (Table I.1B). These are translated to upper distance restraints of 3.5 Å and 4.5 Å, for strong and weak cross peaks, respectively. Sixty-two distance restraints (Table I.1B) were identified from a PDSD spectrum of uniformly 2-^{13}C–labeled fibrils and translated to upper distance limits of 7.5 Å (2-^{13}C glycerol was used as a carbon source (Castellani et al. (2002); LeMaster and Kushlan (1996)) in this sample to reduce spectral overlap, linewidth, and relayed spin-diffusion effects).

2 Materials and Methods

2.1 Preparation of isotope-labeled HET-s(218-289) fibrils.

Isotopically labeled HET-s(218-289) was recombinantly expressed in E. coli, purified and fibrillized as described earlier. HET-s(218-289) with a C-terminal His$_6$ tag was expressed as described for other HET-s constructs (Balguerie et al. (2003)). The bacterial pellets were dissolved and sonicated in 6 M guanidium hydrochloride containing 50 mM TRIS pH 8.0 and 150 mM sodium chloride. The supernatant was cleared by centrifugation for 1 h at 18,000 g. The protein was purified from the supernatant by His-affinity chromatography and concentrated to approximately 0.5-1 mM. Fast buffer exchange was performed to 150 mM acetic acid pH 2.5. Immediately thereafter, the pH was adjusted to 7.5 by addition of 3 M TRIS, which caused HET-s(218-289) to aggregate immediately into amyloid fibrils at 25 °C. The fibrils were washed in H$_2$O and centrifuged into the MAS rotor. At no step, the sample was dried or lyophillized. Extensively labelled samples of HET-s(218-289) were produced using 1,3-^{13}C labeled glycerol and 2-^{13}C labeled glycerol as carbon sources. The expression followed the procedure described in Castellani et al. (2002) and LeMaster and Kushlan (1996). For the production of fibrils consisting of mixtures of differently labelled HET-s(218-289) monomers, the according protein fractions were joined under strongly denaturing conditions (7.5 M GuHCl). All other preparation steps were performed identically for all samples.

2.2 Solid-state NMR experiments.

Two-dimensional NMR experiments were conducted on 14.1 T and 20.0 T (^{1}H resonance frequency 600 MHz and 850 MHz, respectively) wide-bore instruments (Bruker Biospin, Germany) equipped with a 4 mm and a 3.2 mm triple-resonance (^{1}H, ^{13}C, ^{15}N) MAS probes, respectively. All experiments were carried out at probe temperatures of 3 °C to 7 °C. The MAS frequency was set to 19 kHz (20.0 T) and 10 kHz (14.1 T), unless stated otherwise. High-power proton decoupling by SPINAL64 with r.f. amplitudes of 90 kHz–110 kHz was applied during evolution and detection periods.
(i) Experiments on U-[^{13}C, ^{15}N]-HET-s(218-289): For the indirect detection of (^{1}H, ^{1}H) correlations, a CHHC spectrum (Lange et al. (2002)) with a (^{1}H, ^{1}H) mixing time of 200 μs was recorded at 20.0 T. Short contact times of t_{HC} = 200 μs enclosing the (^{1}H, ^{1}H) transfer step favored polarization transfer within bonded

(^{1}H, ^{13}C) pairs only. In addition, an NHHC spectrum with a (^{1}H, ^{1}H) mixing time of 150 µs was recorded at 20.0 T. Contact times of t_{HC} = 200 µs and t_{NH} = 400 µs were used.

(ii) Experiments on U-[^{13}C, ^{15}N] HET-s(218-289) diluted in natural abundance (NA) HET-s(218-289) (1:2.5): A CHHC spectrum with a (^{1}H, ^{1}H) mixing time of 200 µs and t_{HC} = 200 µs was recorded at 14.1 T.

(iii) Experiments on mixed U-^{13}C and U-^{15}N-labeled HET-s(218-289): An NHHC spectrum with a (^{1}H, ^{1}H) mixing time of 150 µs t_{HC} = 200 µs and t_{NH} = 400 µs was recorded at 20.0 T. For this experiment, the MAS frequency was set to 9.5 kHz.

(iv) Experiments on 2-^{13}C glycerol-grown HET-s(218-289): A proton-driven spin diffusion (PDSD) scheme employing a longitudinal mixing time of 250 ms was used. The spectrum was recorded at 14.1 T and 13 kHz MAS.

(v) Experiments on 2-^{13}C glycerol-grown HET-s(218-289) diluted in unlabeled HET-s(218-289) (1:2.5): A PDSD spectrum with a mixing time of 250 ms was recorded at 14.1 T.

(vi) Experiments on 1,3-^{13}C glycerol-grown HET-s(218-289) diluted in unlabeled HET-s(218-289) (1:2.5): A PDSD spectrum with a mixing time of 500 ms was recorded at 20.0 T. Additional experimental parameters and details of the applied labeling schemes are summarized in Table I.1.

All NMR spectra were processed in XwinNMR 3.7 or TopSpin 2.0 (Bruker Biospin) and analyzed using Sparky version 3.113 (T. D. Goddard and D. G. Kneller, University of California, San Francisco) and CARA 1.5 (R. Keller).

2.3 Structure calculation.

For the structure calculation using CYANA (Guntert et al. (1997)), a heptamer was constructed by connecting seven HET-s(218-289) peptide chains with linkers of 41 residues length. All experimental restraints from spectra of diluted samples were classified as intramolecular restraints and were applied for each monomer. The restraints from the undiluted spectra were interpreted as ambiguous restraints (intra- or intermolecular) for the five central monomers, where each distance restraint had to be fulfilled either within the same monomer (intramolecular), or with one of the next neighbours (intermolecular). The upper limit distance restraints were set to 7.5 Å for all ^{13}C–^{13}C contacts and to 3.5 Å or 4.5 Å for the ^{1}H–^{1}H restraints, based on signal intensities. TALOS (Cornilescu et al. (1999)) was used to predict the ψ and ϕ dihedral angles and, where predictions were unambiguous, were added as angle restraints (see Fig. I.1 for details). H-bonds were defined between residues

Spectrum	Sample labelling	^{1}H freq.	MAS	mixing	contact times / µs	TD$_1$	NS	d$_1$	tot. time
PDSD	1,3-^{13}C dil.	850 MHz	19 kHz	500 ms	1000 / – / –	3156	16	1.5 s	25 h
PDSD	2-^{13}C dil.	600 MHz	10 kHz	250 ms	400 / – / –	2048	32	1.0 s	20 h
PDSD	2-^{13}C full	600 MHz	13 kHz	250 ms	1000 / – / –	1280	48	3.0 s	55 h
CHHC	U-[^{13}C, ^{15}N] dil.	600 MHz	10 kHz	200 µs	500 / 200 / 200	1024	512	2.0 s	12 d
CHHC	U-[^{13}C, ^{15}N] full	850 MHz	19 kHz	200 µs	1000 / 200 / 200	1720	128	2.0 s	5 d
NHHC	U-[^{13}C, ^{15}N] full	850 MHz	19 kHz	150 µs	1000 / 400 / 200	768	384	2.0 s	7 d
NHHC	1:1 ^{13}C:^{15}N	850 MHz	9.5 kHz	150 µs	1000 / 400 / 200	320	1536	1.0 s	7 d

Table I.1: Experimental details for the NMR spectra of HET-s(218-289) used for structure elucidation and labelling scheme for the respective sample. 1,3-^{13}C and 2-^{13}C refer to so called extensively labelled compounds as introduced by LeMaster and Kushlan (1996) and Castellani et al. (2002). "full" and "dil." refer to fibrils with either all molecules labelled according to the scheme indicated, or with the labelled molecules diluted in unlabelled (natural isotopic abundance) material (ratio 1:2.5). The mixing times refer to the ^{13}C–^{13}C and ^{1}H–^{1}H spin-diffusion periods for PDSD (Szeverenyi et al. 1982) and CHHC, NHHC (Lange et al. (2002)) spectra, respectively. The contact times refer to the employed cross polarisation steps (Hediger et al. (1995)) (1 or 3, depending on the experimental scheme). "TD$_1$" denotes the number of data points recorded in the indirect dimension, "NS" the number of scans recorded per t_1 experiment, "d$_1$" the inter-scan delay and "tot. time" the total time required to record the spectrum. All spectra were recorded at (278 ± 2) K sample temperature.

which fulfilled all of the following three conditions: (1) their H/D exchange rate was found to be slow ($\log(k_{ex}\,h^{-1}) < -1$ (Ritter et al. (2005))), (2) TALOS gave a secondary structure prediction in β-sheet conformation and (3) peaks connecting H_i^N and $H_{i\pm36}^\alpha$ of adjacent molecules were observed in the NHHC spectrum of uniformly labelled material (Fig. I.4D).

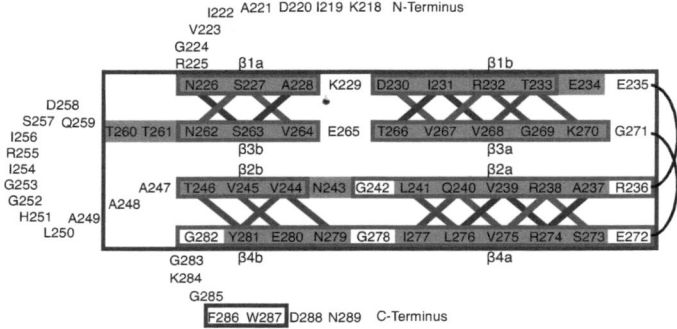

Figure I.1: Summary of the TALOS, symmetry and H-bond restraints used in the structure calculation of the HET-s(218-289) heptamer. The residues for which TALOS (Cornilescu et al. (1999)) predicted a β-sheet conformation are underlaid by green areas. Residues which show slow H/D exchange are indicated by red boxes (Ritter et al. (2005)). Residues having both properties were defined to be hydrogen bonded if the corresponding NMR peaks were present in the NHHC spectra. Coloured lines between residues designate the H-bonds used in the structure calculation. Red, blue and violet lines label intramolecular, intermolecular and ambiguous inter-/intramolecular restraints, respectively. The blue boxes enclose all assigned residues, for which symmetry restraints were added in the calculation.

Information about the intra- or intermolecular character of the H-bonds was obtained from the NHHC spectrum of the 1:1 ^{13}C:^{15}N labelled sample. Signals appearing in this spectrum clearly identify the intermolecular character of 7 H-bonds as well as the intramolecular character of 7 further H-bonds. The intra-/ intermolecular character of the other H-bonds remained unclear due to spectral overlap and they were implemented as ambiguous intra-/intermolecular restraints in CYANA (see Fig. I.1). As described in the main text, it is known, that the fibrils are stacked as a quasi-one-dimensional array of molecules along the fibril axis. As only a single set of resonances is observed in the NMR spectra, all molecules in the fibril must be structurally equivalent. This leads to a total of 206 distance restraints for the CYANA calculation, introduced as lower and upper distance restraints of 9.2 Å and 10.0 Å, respectively. All restraints in the calculation were used with the same weight.

A complete list of the structural restraints obtained by NMR and used in the structure calculations is given in Table I.1. For the initial structure calculations, only the spectrally unambiguous peaks (Table I.1A) were used. Unambiguous peaks, in this context, refer to cross-peaks between resonances which both have only a single assignment possibility within a spectral window of ± 0.15 ppm. Based on the initial structure, a final set of unambiguous restraints was picked from the spectra. For these peaks, only one of the possible assignments was compatible with the initial structure, meaning that all but one assignment (from spectral properties) could be excluded by criterion that all other possible partners were at a distance of more than 12 Å, as judged by the initial structure. These restraints are listed in Table I.1B and depicted in Fig. I.2 and were used as input for the final structure calculation.

The residue-residue plot corresponding to the peak-lists (Fig. I.5) allows for the identification of a typical parallel β-sheet pattern. We found 69 peaks connecting residues i and j with $35 \leq |i - j| \leq 37$, supporting a parallel beta-sheet between the two pseudo-repeats (res. 226-246 and res. 262-282), which are separated by 36 residues in the poly-peptide sequence. Additionally, the plot shows that the set of unambiguous peaks (lower right half) already contains the elements of the plot for the final set of restraints (upper left half). This is also reflected in the fact that the structure calculations for both sets lead to closely related folds of the HET-s(218-289) fibrils (data not shown).

In total, 200 calculations were performed and the 5 central molecules of the 20 lowest energy structures were further analysed. The central molecules of the structures in this ensemble are displayed in Fig. I.3. The structural statistics are given in Table S3. For the ensemble of the 20 lowest energy structures, the average root mean square distances (rmsd) is 1.16 Å for all assigned heavy atoms, showing that a single fold emerges from the structure calculation. The Ramachandran analysis accomplished by Procheck NMR (Laskowski et al. (1996)) shows that no residue of the lowest energy structure adopts disallowed conformations regarding ϕ and/or ψ dihedral angles. The values for the ensemble of the 20 lowest energy structures are only slightly worse. The obtained structures were visualized using PyMol 0.99 (DeLano Scientific LLC) and MolMol 2K.1 (Reto Koradi, Institut fuer Molekularbiologie und Biophysik, ETH Zurich).

I The Structure of HET-s(218-289)

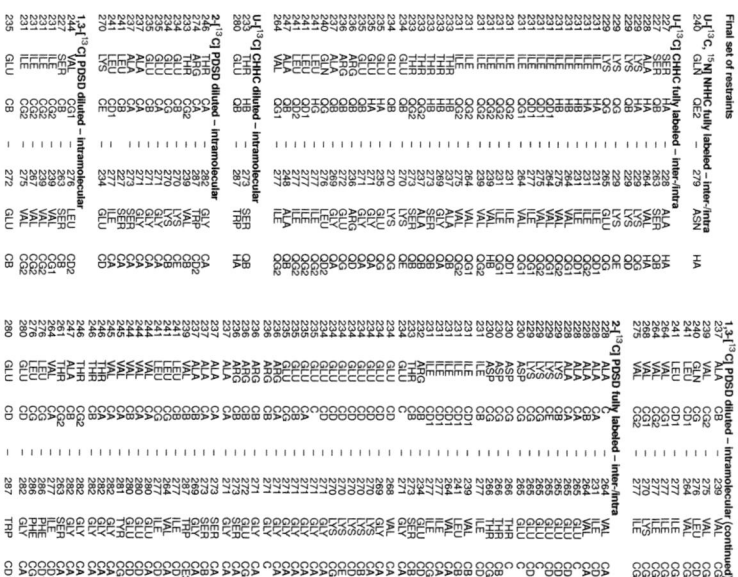

Table I.1: Caption on following page.

Table I.1: (previous page) Summary of distance restraints as obtained from the NMR spectra and used in the structure calculations. (**A**) Unambiguous peaks used for the first round of calculations and, (**B**) final set of restraints as used in the final structure calculation. Note that all restraints from (**A**) are included in table (**B**).

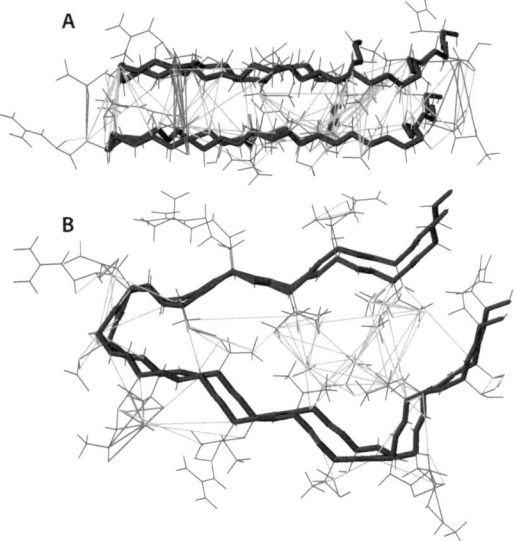

Figure I.2: The distance restraints from Table. I.1B are displayed for the hydrophobic core residues. Side chain - side chain constraints supporting salt-bridges are colored red. (**A**) Side-view, (**B**) top-view of the central monomer of the lowest-energy structure. Note that intra- as well as ambiguous inter-/intramolecular restraints are shown in one HET-s molecule (unlike their implementation in the structure calculation).

I   The Structure of HET-s(218-289)

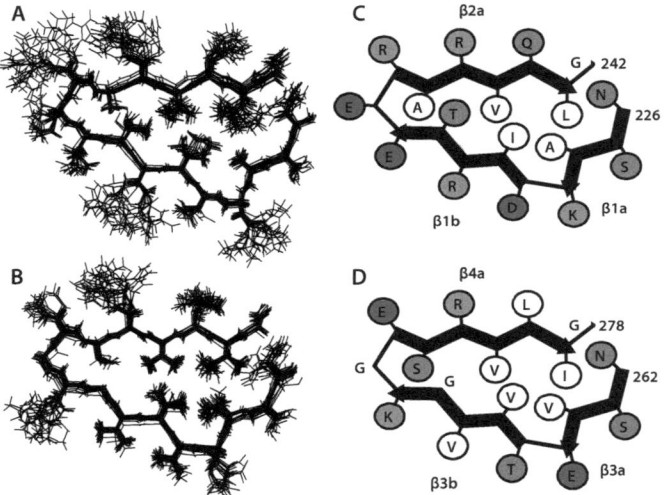

Figure I.3: (**A**) Lower and (**B**) upper layer of the bundle of the 20 lowest-energy structures for the hydrophobic core of the central molecule. The structures were superimposed on the backbone atoms of (N226-G242) and (N262-G278). (**C** and **D**) Schematic representations of the two layers.

A	**Average rmsd to mean coordinates**	Backbone heavy atoms / Å	All heavy atoms / Å
	S226–G242, S262–G278 (see Fig. 3)	0.37 ± 0.11	0.99 ± 0.11
	S226–A248, T260–G282, F286, W287 †	0.63 ± 0.13	1.16 ± 0.09
	† Extent of resonance assignment		

B	**Ramachandran analysis**	Lowest energy structure	20 lowest energy structures
	most favoured regions	71.7 %	69.1 %
	additional allowed regions	26.7 %	24.9 %
	generously allowed regions	1.7 %	4.3 %
	disallowed regions	0.0 %	1.7 %

C	**Structure calculation statistics**	Lowest energy structure	20 lowest energy structures
	Residual CYANA target function (heptamer) / Å^2	1.80	5.44 ± 2.07
	Residual distance restraint violations* ≥ 0.2 Å	0	1 (max. 0.74 Å)**
	Residual dihedral angle restraint violations* ≥ 5°	0	0**
	Residual van der Waals violations* ≥ 0.2 Å	0	1 (max. 0.24 Å)**

* For the five central monomers ** restraints violated in six or more structures

Table I.2: Structural statistics for the HET-s(218-289) heptamer. The averages were performed for the 20 lowest energy structures out of a total of 200 calculated structures. (**A**) Average rmsd to the mean coordinates calculated for the central molecule for the hydrophobic core only and for all residues. (**B**) Results of the Ramachandran analysis of the central molecule for the non-Gly residues of HET-s(218-289) by Procheck NMR (Laskowski et al. (1996)). (**C**) Structure calculation statistics: The residual CYANA target function for the whole heptamer and the violations per molecule for the five central HET-s(218-289) molecules are given.

I The Structure of HET-s(218-289)

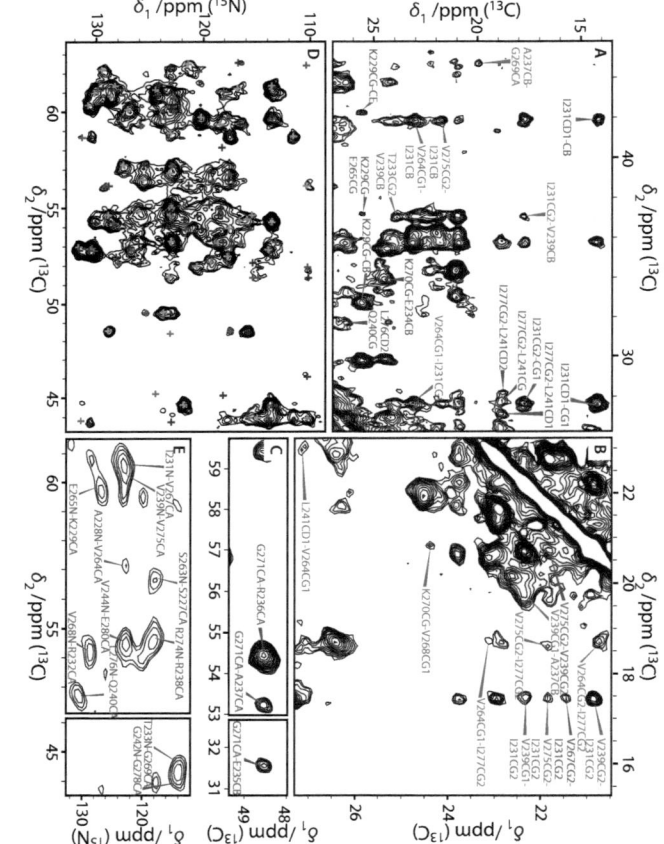

Figure I.4: Caption on following page.

Figure I.4: (previous page) Solid-state NMR spectra of HET-s(218-289) fibrils. Blue labels identify signals that correspond to short-range restraints ($|i - j| \leq 1$); red labels, signals that correspond to register-restraints ($35 \leq |i - j| \leq 37$); and green labels, restraints that define the hydrophobic core. **A** CHHC spectrum of U-[^{13}C, ^{15}N]-labeled HET-s(218-289) fibrils recorded at 19 kHz MAS with 150 μs ^{1}H–^{1}H mixing with register and side-chain contacts useful for structure calculation. **B** PDSD spectrum of fibrils containing 25% 2-^{13}C-labeled monomers diluted in unlabeled HET-s(218-289), recorded at 10 kHz MAS with 250 ms ^{13}C–^{13}C mixing giving intramolecular contacts. **C** PDSD spectrum of fibrils containing 25% 1,3-^{13}C-labeled molecules diluted in unlabeled HET-s(218-289), recorded at 19 kHz MAS with 500 ms ^{13}C–^{13}C mixing giving further intramolecular contacts. **D** NHHC spectrum of a U-[^{13}C, ^{15}N]–labeled sample recorded at 19 kHz MAS with 150 μs ^{1}H–^{1}H mixing. The expected peak positions calculated for proton pairs separated by less than 3.5 Å in the lowest-energy structure are indicated by colored crosses. For a β sheet, three types of $H_i^N - H_j^\alpha$ contacts are mainly expected: intraresidue ($j = i$), sequential ($j = i - 1$), and interstrand register ($|j - i| = 36$, intra- and intermolecular) peaks are indicated with blue, orange, and red symbols, respectively. The green crosses indicate the positions of other expected peaks. **E** NHHC spectrum of fibrils containing a 1:1 mixture of U-^{13}C- and U-^{15}N-labeled monomers, recorded at 9.5 kHz MAS with 150 μs ^{1}H–^{1}H mixing. Only intermolecular proton contacts give rise to peaks in this spectrum.

3 Spectra and Distance Restraints

In fibrils with uniform isotope labeling, distance restraints measured by NMR are difficult to assign to either intra- or intermolecular contacts. A comparison of spectra from uniformly labeled samples with those from "diluted samples", in which isotopically labeled monomers are mixed with unlabeled material before fibrilization (ratio 1:2.5), can in some cases resolve the ambiguities. Following this strategy, we identified a total of 30 intramolecular restraints from a CHHC spectrum on a diluted U-[^{13}C, ^{15}N]–labeled HET-s(218-289) sample and PDSD spectra on diluted extensively labeled HET-s(218-289) (Fig. I.4, B and C) (Castellani et al. (2002); LeMaster and Kushlan (1996)). Intermolecular restraints were obtained from a sample fibrilized from a mixture of U-^{13}C and U-^{15}N labeled HET-s(218-289) molecules (Etzkorn et al. (2004)). Polarization transfer between ^{15}N- and ^{13}C-bound protons, respectively (Lange et al. (2002)) (Fig. I.4E), indicates short H^N–H$^\alpha$ contacts (3.0 Å) between β-sheets of different monomers and thus selectively characterizes the intermolecular interface. Comparing these spectra to similar spectra on uniformly labeled compounds allowed us to identify seven H-bonds as intramolecular and seven H-bonds as intermolecular, and nine H-bonds remained ambiguous (intra- or intermolecular). Details are given in the Materials and Methods section. These H-bonds define a parallel in-register arrangement of the intra- and intermolecular β-sheet interfaces, which is supported by 69 of the experimental restraints that correlate residues i and $i+(36\pm1)$ (Fig. I.4 and Fig. I.5). Fourteen of these can be identified unequivocally as intramolecular restraints.

All restraints used for the structure calculation are summarized in Table I.3 (A comprehensive list of cross peaks is given in Table I.1 in the appendix). In total, 90 ^{13}C–^{13}C and 44 ^{1}H–^{1}H distance restraints (i.e., 2.8 per assigned residue, all of them containing nontrivial structural information) were identified, together with the 23 β-sheet H-bonds (see Fig. I.1 in the appendix). In addition, 74 angle restraints obtained by TALOS (Cornilescu et al. (1999)) were used (Fig. I.1). From NMR (Ritter et al. (2005)) and mass-per-length measurements (Sen et al. (2007)), it is known that the thinnest HET-s(218-289) fibril consists of a stack of single molecules all having the same structure. To implement the resulting quasi–one-dimensional (1D) symmetry, we used 206 additional intermolecular distance restraints for the sequentially assigned residues (Fig. I.1).

The NMR-structure calculation was conducted with CYANA (Guntert et al. (1997)) using the restraints of Table I.3 on a set of seven molecules, and yielded the struc-

Spectra and Distance Restraints 3

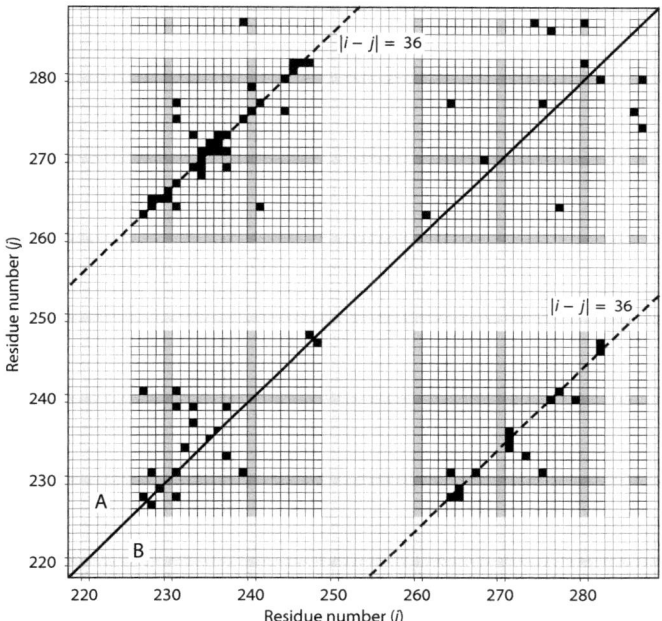

Figure I.5: Residue-residue plot with black squares indicating pairs of residues connected by at least one distance restraint. In the upper left half (**A**), all NMR restraints used in the final structure calculation are shown, the lower right (**B**) shows only restraints from unambiguous peaks (see text), as used in the first round of calculations. The dotted red lines indicate positions where peaks supporting in-register parallel β-sheets as obtained in the final structure are expected. For unassigned residues the grid is shaded; the grey strips serve solely for orientational purposes.

		total	^1H–^1H	^{13}C–^{13}C	unamb. intramol.
distance restraints	short-range ($\|i-j\| \leq 1$)	12	12	–	–
	med.-range ($2 \leq \|i-j\| \leq 4$)	9	2	7	3
	register ($35 \leq \|i-j\| \leq 37$)	68	13	55	14
	long-range (other)	45	17	28	13
	total	134	44	90	30

type of restraint	total	intramol.	intermol.	amb.
H-bond	23	7	7	9
periodicity	206	–	206	–
dihedral angles (TALOS)	74			

Table I.3: Number of structural restraints used for the structure calculation (per molecule).

ture depicted in Fig. I.6. The backbone heavy-atom average root mean square deviation to the mean structure of the 20 lowest-energy conformers (200 calculated structures) is 0.4 Å for the backbone and 1.0 Å for all heavy atoms, considering only the rigid core of one HET-s(218-289) molecule (residues N226 to G242, N262 to G278) (Fig. I.6) and Fig. I.3).

4 The Structure of the Hydrophobic Core

The overall organization of a HET-s(218-289) fibril is a left-handed β-solenoid with two windings per molecule (Fig. I.6A). The core of the fibril is defined by three β-strands per winding (six β-strands per molecule) that form continuous in-register parallel β-sheets. An additional β-sheet outside the core is formed by β_{2b} and β_{4b} (Fig. I.6B). This organization is consistent with the fold proposed earlier (Ritter et al. (2005)), but the details show that each previously proposed β-strand is split into two shorter segments (e.g., β_1 into β_{1a} and β_{1b}). The segments β_{1a} and β_{1b} (β_{3a} and β_{3b}) are connected by a two-residue β-arc, changing the inside-outside pattern of side chains and leading to an approximately rectangular kink in the strand at K229 and E265, for β_1 and β_3, respectively (Fig. I.6B). The connection between β_{1b} and β_{2a} (and similarly, between β_{3b} and β_{4a}) is provided by a three-residue β-arc, allowing for the orientation change of the polypeptide backbone by 150°. A disruption of the β-sheet pattern is also observed between β_{2a} and β_{2b} (β_{4a} and β_{4b}), leading to 90° β-arcs. β_1-β_2 and β_3-β_4 are pseudo-repeats and form both parallel intramolecular and intermolecular H bonds, as follows: β_{1a}-β_{3a}, β_{1b}-β_{3b}, β_{2a}-β_{4a}, β_{2b}-β_{4b} (Fig. I.6, A and D).

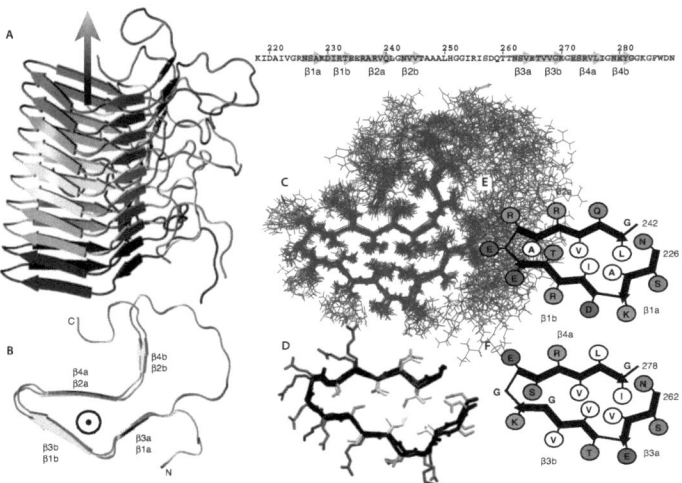

Figure I.6: Structure of the HET-s(218-289) fibrils. The fibril axis is indicated by an arrow. **A** Side view of the five central molecules of the lowest-energy structure of the HET-s(218-289) heptamer calculated from the NMR restraints. **B** Top view of the central molecule from **A**. β_3 and β_4 lie on top of β_1 and β_2, respectively. **C** NMR bundle: superposition on residues N226 to G242, N262 to G278 of the 20 lowest-energy structures of a total of 200 calculated HET-s(218-289) structures. Only the central molecule of the heptamer is shown. **D** Representation of the well-defined central core of the fibril (N226 to G242, N262 to G278). Hydrophobic residues are colored white, acidic residues red, basic residues blue, and others green (lowest-energy structure). (**E** and **F**) Schematic representations of the two windings in **D**: the first winding (N226 to G242, displayed in **E**) of the β solenoid is located beneath the second one (N262 to G278, displayed in **F**).

As seen in Fig. I.6, D to F, the β-sheet arrangement is stabilized by favorable side-chain contacts. The first three β-strands of each pseudo-repeat enclose a triangular hydrophobic core that is tightly packed and contains almost exclusively hydrophobic residues (Ala, Leu, Ile, and Val) with numerous experimental restraints (indicated by green labels in the spectra of Fig. I.4) between hydrophobic side chains.

39

The packing is dense and defines a dry interface. The only polar residues in the core are T233 and S273, which can form side-chain H bonds, thereby stabilizing the formation of the turn between β_1 and β_2. In contrast, all charged residues face outside and are mostly located in β-arc regions, where the solvent accessibility is high. Three of them are arranged on top of each other such that charges are compensated (Fig. I.6, D to F) and the formation of salt-bridges becomes possible. Several experimental restraints support the existence of the salt-bridges K229-E265, E234-K270, and R236-E272. (Fig. I.2). Because the stacking is parallel, we expect the charge compensation to have both intra- and intermolecular character. This may explain why fibrils have high stability against denaturation by nonionic urea at neutral pH, but are destabilized by urea at acidic or basic pH (Sabate et al. (2007)). The two asparagines next to the hydrophobic core are stacked and can form a ladder (N226-N262), further contributing to the fibril stability through side-chain H-bonds. Another asparagine ladder can be formed outside the hydrophobic core (N243-N279) (Yoder et al. (1993)).

For the structure calculation, only a limited number of well-resolved NMR peaks was used. To assess whether the structure is consistent with all the peaks observed in the spectra, we calculated the expected cross peaks from the internuclear distances in the structure. As an example, all expected resonances in the NHHC spectrum are shown as symbols in Fig. I.4 D. Similar good agreement was found for the other spectra (see supporting information of Wasmer et al. (2008a) and Van Melckebeke et al. (2010)). In our structure, the cross-section of the fibril is approximately circular (Fig. I.6C), a feature which might explain why no helical twist of HET-s(218-289) fibrils has been detected in electron micrographs (Sen et al. (2007)). The structure given in Fig. I.6 explains all the details of the chemical-shift data, H/D exchange, water accessibility, and mutant studies by Ritter et al. (Ritter et al. (2005)), even the ones that could not be explained by the earlier straight-stranded fold. The non–β-sheet chemical shift of K229 and E265 residues, together with their fast H/D exchange rates, can now be explained by the β-arc at this position. Residues D230 and T266, which were found to be highly solvent-accessible as probed by chemical cross-linking of cysteine mutants (Ritter et al. (2005)), are indeed solvent-exposed in our structure (Fig. I.6, D to F), whereas I231, T233, V239, L241, V267, and V275 are buried in the hydrophobic core of the fibrils, explaining their very low water accessibility. G242, N279 and E280, for which intermediate values were found, are located in a region of β_{2b} and β_{4b} where the NMR structure is less well defined and where both sides of the β-strand are somewhat protected (Fig. I.6C). The present model also explains the requirement for a minimal length of the loop connecting β_{2b} and β_{3a} to maintain infectivity (Ritter et al. (2005)).

The structure of HET-s(218-289) shows an overall β-helical fold that is of higher structural complexity than that of peptide fibrils (Nelson and Eisenberg (2006); Sawaya et al. (2007))—a complexity that is reminiscent of soluble protein folds. Part of the complexity, e.g., the favorable alternation of positive and negative charges in ladders along the fibril axis, can be realized only because of the pseudo-repeat (β_1/β_3 and β_2/β_4), leading to a structure in which one molecule forms two turns of the solenoid. This feature distinguishes HET-s from other amyloids and prions that have also been modeled by β-solenoids (Kajava et al. (2006); Lazo and Downing (1998)). It leads to the formation of three salt-bridges that stabilize the structure, in contrast to the finding in yeast prion protein Sup35 where the presence of pseudo-repeats is probably related to structural variability and the existence of prion strains (Toyama et al. (2007)). The three-stranded triangular hydrophobic core indeed bears some resemblance to β-solenoid structures of soluble proteins like filamentous hemagglutinin (Clantin et al. (2004)) and the P22 tailspike protein (Steinbacher et al. (1996)). In contrast to HET-s(218-289), these structures are not periodic, but the geometry of the triangular core is quite similar. Furthermore, a β-solenoid fold has also been proposed for the prion state of the human prion protein PrP on the basis of modeling and electron microscopy by Govaerts et al. (2004) and for the yeast prion Sup35 by Krishnan and Lindquist (2005) and Kishimoto et al. (2004).

The well-organized structure of the HET-s prion fibrils can explain the extraordinarily high order in these fibrils, as seen by NMR, as well as the absence of polymorphism caused by different underlying molecular structures at physiological pH conditions, because the specific nature of the interactions in the fibril excludes polymorphic molecular conformations with comparable stability. The fibril structure of HET-s(218-289) exemplifies the well-defined structure of a functional amyloid and illustrates the interactions that can stabilize their fold (Fowler et al. (2007)).

5 Back-calculations

To confirm the validity of the structure, we performed back-calculations of the positions of expected peaks from our final structure. The distances between these pairs were extracted from the structure and the corresponding peaks plotted onto the spectra. Figs. I.7 and I.8 show the back-calculations for the intermolecular NHHC spectrum and the CHHC spectrum, respectively.

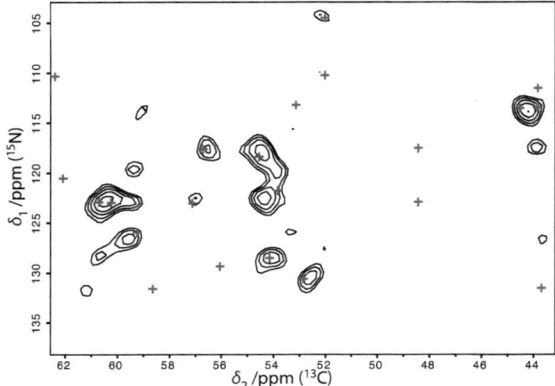

Figure I.7: NHHC spectrum of fibrils containing a 1:1 mixture of U-[^{13}C] and U-[^{15}N] labelled monomers recorded at 9.5 kHz MAS with 150 µs ^{1}H–^{1}H mixing (same spectrum as in Fig. I.4E). A back-calculation was performed for the lowest energy structure for proton pairs connecting neighbouring HET-s(218-289) molecules (i.e. intermolecular contacts) and separated by less than 3.5 Å. Red crosses are drawn for atoms belonging to residues i and $i + 36$ (of different molecules), other correlations are labelled green. Note that the three two-contours peaks without crosses can be explained by inter-atomic distances only slightly larger than 3.5 Å.

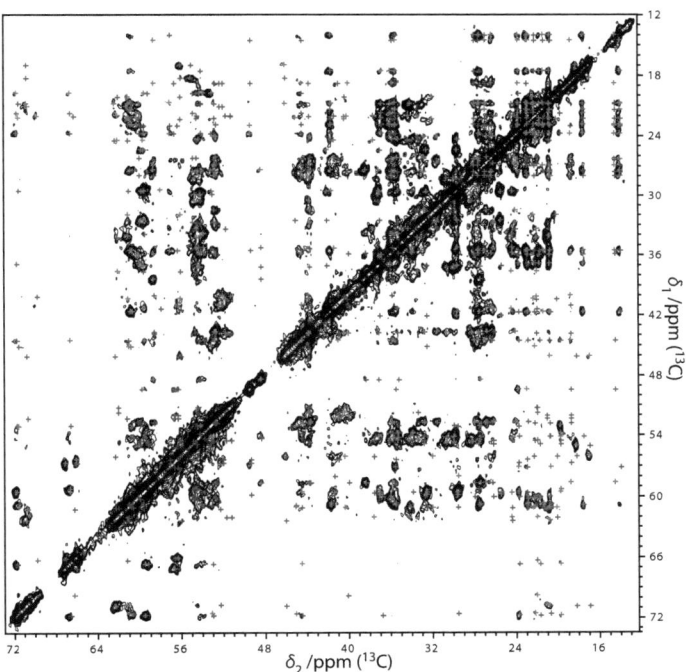

Figure I.8: CHHC spectrum of U-[^{13}C, ^{15}N] fully labelled HET-s(218-289) recorded at 19 kHz MAS with 150 μs ^{1}H–^{1}H mixing (same spectrum as in Fig. I.4A). A back-calculation was performed for the lowest energy structure for proton pairs separated by less than 3.5 Å (red crosses) and up to 4.5 Å (blue crosses).

I The Structure of HET-s(218-289)

Chapter II

Full-length HET-s

This work was done in collaboration with Anne Schütz, Antoine Loquet, Caroline Buhtz and Jason Greenwald and is published (Wasmer et al. (2009b)). Anne performed the experiments on HET-s(1-227), Antoine produced the molecular model (Fig. II.9) and Caroline and Jason provided support for the sample preparation.

1 Introduction

Prions are associated with transmissible traits in which the agent of transmission is a conformationally altered form of the "properly" folded protein. For the mammalian prion diseases (Prusiner (1982); Wells et al. (1987); Will et al. (1996); Collinge (2001); Mathiason et al. (2006); Wadsworth et al. (2008)), the prion form is believed to be associated with elongated insoluble aggregates termed fibrils. Similar observations have been made for yeast and fungal prions for which the phenotypes [URE3], [PSI], [PIN] and [Het-s] have been shown to be intimately linked to the prion form of the proteins Ure2, Sup35, Rnq1, and HET-s, respectively. The HET-s prion studied here can exist as a soluble protein, while also being able to go into a fibrillar prion state (Coustou et al. (1997)). In the soluble state, the C-terminal residues 228-289 are flexibly disordered in vitro, while the N-terminal part, approximately residues 1-227, forms a mainly α-helical globular domain (Balguerie et al. (2003)). In the prion form, a proteinase-K resistant fragment appears, consisting of residues 218-289 and this is referred to as the prion domain (note that it partially overlaps with the globular domain 1-227). The het-s/S locus in the filamentous fungus *Podospora anserina* has two naturally occurring alleles, het-s and het-S, that encode for two proteins, each comprising 289 amino acid residues and differing in only 13 residues that are evenly distributed over the sequence (Turcq et al. (1991)). The het-S allele encodes for the HET-S protein, and although it also contains a prion domain, it has so far been found exclusively in a soluble, non-prion state. Fungal cells bearing the het-s allele can display one of two phenotypes: those with the non-prion phenotype [Het-s*] behave neutrally towards cells bearing the het-S allele; those displaying the [Het-s] phenotype, believed to contain the aggregated prion form of HET-s, display an antagonistic behavior when confronted with a het-S strain. The resulting cell death-like reaction is termed heterokaryon incompatibility. The prion domain of HET-s, HET-s(218-289), is necessary and sufficient for prion infectivity (Rizet (1952); Saupe (2000)). This prion domain alone forms fibrils in vitro (Dos Reis et al. (2002)) and can, as a GFP-fusion protein, propagate the prion form in vivo (Balguerie et al. (2003)). The globular domain also plays an important role, as it alone determines the phenotype of fungi harboring chimeras of HET-s and HET-S. Although HET-s and HET-S differ in both their N-terminal and C-terminal domains (by 10 and 3 residues, respectively (Balguerie et al. (2003))), it is the differences in the N-terminal domain that determine the role of the protein in the incompatibility reaction (Dalstra et al. (2005)). Despite the tremendous importance of the 3D structure in the context of the prion hypothesis, there are no atomic resolution structures for the fibril form of a full-length prion protein, and HET-s(218-289) remains the only prion for which an atomic resolution structure in the fibrillar form is known (Wasmer et al. (2008a)). There are however structural models for full-length prions which postulate that the corresponding fibrils contain an amyloid-like spine, formed by the prion domain, to which the globular domain is attached via a flexible linker (Wickner et al. (2008)). These models usually invoke an ordered prion domain on which the globular domains are irregularly ordered so that there is no symmetry element relating one globular domain to the others. There is, however, also experimental evidence that prion fibrils can be devoid of a cross-β-core as in the case of the Ure2 protein (Bousset, Thomson et al. 2002). For the isolated prion domains of all known yeast prions, there is ample evidence that the fibrils indeed contain a cross-β spine as the dominant structural element(Wickner, Shewmaker et al. 2008). The recently reported structure of HET-s(218-289) shows that the fibrils are formed by a highly-ordered triangular core formed by three parallel cross-β sheets (Wasmer et al. (2008a)). The finding that the hydrogen/deuterium exchange rates in the isolated prion domain HET-s(218-289) are very similar to the exchange rates of the protons in the prion domain of (full-length) HET-s (Ritter et al. (2005)) suggested early-on that the structure of the isolated prion domain is basically conserved in the context of amyloids of full-length HET-s.

Here we present a structural study of the full-length HET-s prion in its fibrillar form. The modular structure of the protein allowed us to use a strategy that is based on a comparison of the following three proteins by solid-state NMR: (i) fibrillized full-length HET-s, (ii) the isolated prion-domain HET-s(218-289) in its amyloid form and (iii) the N-terminal domain HET-s(1-227) in a crystalline form. Note that both of the HET-s fragments contain residues 218-227 since these residues belong to the HET-s globular domain when in the soluble, non-prion form, but are also involved in the fibrils formed by the prion domain alone. Indeed, HET-s(1-227) has previously been shown by liquid-state NMR to be well-structured in solution and to display the same mostly α-helical fold as in the soluble form of the full-length

II Full-length HET-s

HET-s protein (Balguerie et al. (2003)). Here we demonstrate that HET-s(1-227) can be crystallized in a form that yields highly resolved solid-state NMR spectra. By comparing the spectra recorded on the isolated domains to the full-length protein, we show that the prion domain has virtually the same structure in the fibrils of HET-s and HET-s(218-289), while the N-terminal domain is disordered in the fibrils of full-length HET-s, displaying an enhanced flexibility and features reminiscent of a molten globule.

2 Methods

2.1 Sample preparation

U-[^{13}C, ^{15}N] labeled HET-s-His$_6$ was recombinantly expressed in *E. coli* in inclusion bodies. Cells were lysed and centrifuged at 18,000 g, 4 °C for 10 min. The pellet was resuspended in 7.5 M GuHCl, 150 mM NaCl, 50 mM TrisHCl, pH 7.4 and incubated for 18 h at 50 °C. The supernatant of this solution (after centrifugation for 1 h at 250,000 g) was loaded onto a Nickel column (HisTrap FF, GE Healthcare) and HET-s-His$_6$ was eluted using 8 M Urea, 150 mM NaCl, 50 mM Tris·HCl, pH 7.4, 200 mM Imidazole. Thereafter, Dithiothreitol (DTT) was added to a concentration of 20 mM and the buffer was exchanged to 150 mM NaCl, 150 mM TrisHCl, pH 8.0, 1 mM DTT on a HiTrap Desalting 5 ml column (GE Healthcare). In this buffer, the sample refolded (judged from the NMR spectrum given in the Appendix, Fig. X.5) and finally fibrillized spontaneously. In total, 3 samples were prepared in this way, 2 of them fibrillized at a concentration of about 1 mg/ml (one of them with seeding with HET-s fibrils) and one at 0.5 mg/ml. In the end, the salt was largely removed by dialyzing against pure water for one high-concentration and the low-concentration sample. The fibrils were centrifuged into a 3.2 mm Bruker MAS rotor at 200,000 g. All three samples yielded very similar NMR spectra. All spectra displayed herein were taken with the salt-free sample fibrillized at 1 mg/ml without seeding.

HET-s(1-227) crystals were prepared starting from 34 mg of protein in a 17 mg/ml protein solution in 20 mM Tris HCl pH 8.5, 0.5 mM DTT and 0.02 % NaN$_3$. Crystals were grown in sitting drops of 150 ml by adding 150 ml of precipitation solution containing 30 % PEG 4000 in 20 mM Tris HCl pH 8.5. The reservoir contained 20 ml of a 2 M NaCl solution. Abundant precipitates appeared within one week and were harvested and filled into a 3.2 mm rotor using a centrifuge at 5000 g, followed by a final centrifugation at 84,200 g for 30 minutes in an ultracentrifuge. HET-s(218-289) fibrils were prepared as described in references (Balguerie et al. (2003); Siemer et al. (2006b)). Basically, the procedure is the same as for HET-s except that the addition of DTT can be omitted, as HET-s(218-289) does not contain a cysteine.

2.2 Solid-State NMR

All spectra were recorded on a Bruker Avance II+ 850 spectrometer operating at a static magnetic field of 20.0 T using a 3.2 mm Bruker triple resonance probe. For HET-s and HET-s(1-227), a 3.2 mm Bruker triple resonance probe equipped with an LLC coil was used to reduce r.f. heating during the experiments. The mixing period for ^{13}C–^{13}C 2D correlation spectra was set to 100 ms and the ^{1}H r.f. field strength was optimized to obtain the most effective transfer from aliphatic carbons to carbonyl carbons (which was found to be neither the exact MIRROR (Scholz et al. (2008)) nor DARR (Takegoshi et al. (2001)) condition). The DREAM (Verel et al. (2001)) mixing period in the NCACB was 4 ms. The spectra were recorded at 19 kHz (HET-s(218-289)), 18 kHz (HET-s) and 17 kHz HET-s(1-227)) MAS at sample temperatures of about 3 °C and 100 kHz SPINAL-64 ^{1}H-decoupling during t_1, t_2 and t_3 evolution periods. The maximal t_1 and t_2 acquisition times were 14 ms for all DARR and NCA spectra and 7 ms for the NCACB spectrum. Maximal t_3 was 14 ms for the NCACB spectrum.

Carbon-detected INEPT (Morris and Freeman (1979); Burum and Ernst (1980); Andronesi et al. (2005)) experiments were recorded at 20 °C sample temperature and 50 kHz SPINAL-64 ^{1}H-decoupling during t_2 but otherwise identical conditions.

All spectra were processed in TopSpin or XwinNMR (Bruker Biospin) and analyzed using Sparky version 3.113 (T. D. Goddard and D. G. Kneller, University of California, San Francisco).

3 Results

To characterize the full-length HET-s fibrils, we compared their NMR spectra to the corresponding spectra of HET-s(218-289) in its fibrillar form and of HET-s(1-227) in a crystalline form. Earlier measurements on the amyloids of HET-s(218-289) showed that it consists of both rigid and highly dynamic elements (Siemer et al.

(2006b)). To address the rigid part of the three proteins, cross-polarization (CP) (Hartmann and Hahn (1962); Pines et al. (1973); Hediger et al. (1994)) based NMR experiments were performed. To identify and characterize the dynamic (flexible) parts, carbon-detected experiments employing an INEPT transfer step (Morris and Freeman (1979); Burum and Ernst (1980); Andronesi et al. (2005)) instead of cross polarization were employed.

Spectra of the the prion-domain HET-s(218-289) in its amyloid form have been described (Siemer et al. (2005, 2006b,a); Wasmer et al. (2008a)). HET-s(1-227) was obtained in its crystalline form as described in the materials and methods section. Fibrils of HET-s were formed from protein extracted from E. coli inclusion bodies. Care was taken to ensure that the globular domain was folded by comparing the liquid-state NMR spectra to that of the isolated globular domain (Balguerie et al. (2003)) prior to fibrillization (Appendix, Fig. X.5).

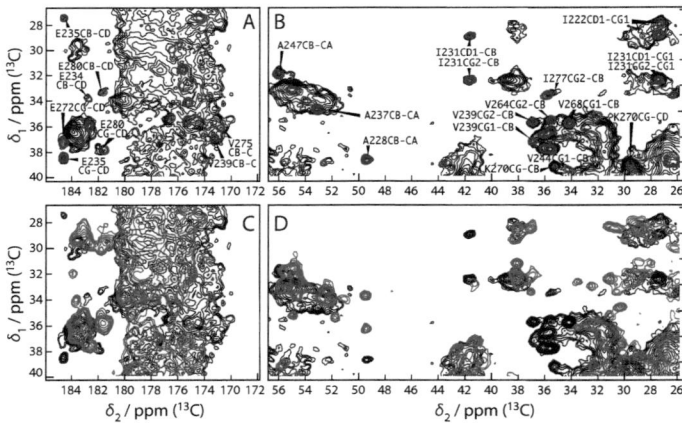

Figure II.1: A, B Extracts of the 100 ms DARR spectra of HET-s(218-289) (blue) and HET-s (black). All peaks of the prion domain are present in the spectrum of the HET-s fibrils. **C, D** Extracts of the 100 ms DARR spectra of HET-s(1-227) (red) and HET-s (black). For the full spectra see Figs. X.6 and X.7 in the Appendix. All peaks observed for the prion domain are present in the spectrum of full-length HET-s fibrils.

The prion domain has the same structure in isolation as in the full-length HET-s fibrils. A superposition of extracts from 2D ^{13}C–^{13}C DARR (Takegoshi et al.

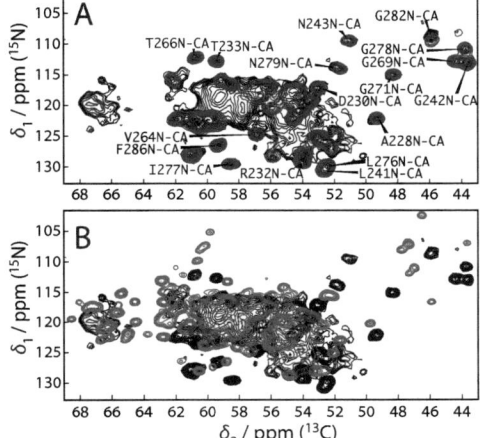

Figure II.2: A NCA correlation spectra of HET-s(218-289) (blue) and HET-s (black). **B** NCA correlation spectra of HET-s(1-227) (red) and HET-s (black).

(2001)) and ^{15}N–^{13}C NCA (Baldus and Meier (1996)) spectra of HET-s fibrils and HET-s(218-289) fibrils is shown in Fig. II.1A, B and Fig. II.2A, respectively (full spectra are given in Fig. X.6). The HET-s spectrum (black) globally appears not as well resolved as the spectrum of HET-s(218-289) (blue); it however contains all blue signals (with very few exceptions, see below) arising from HET-s(218-289), and shows comparable linewidths (FWHH 95 Hz - 140 Hz for HET-s vs. 85 Hz - 120 Hz for HET-s(218-289)). The additionally observed resonances (black signals where there are no blue signals), which are expected to come mainly from residues 1-217 of HET-s, are significantly less well resolved. The NCA spectra in Fig. II.2A also reveal several well resolved N–C$^\alpha$ cross-peaks. Again, all peaks of the blue spectrum representing the prion domain are as well visible in the black HET-s spectrum, confirming the findings from the ^{13}C–^{13}C correlation spectra.

To reduce spectral overlap, a three-dimensional NCACB spectrum (Hong (1999))

Table II.1: Comprehensive list of chemical shifts for residues observed in the CP spectra (C–C DARR, NCA and NCACB) of the prion domain, residues 218-289 of HET-s. For resonance frequencies determined by multiple peaks, the average value is given.

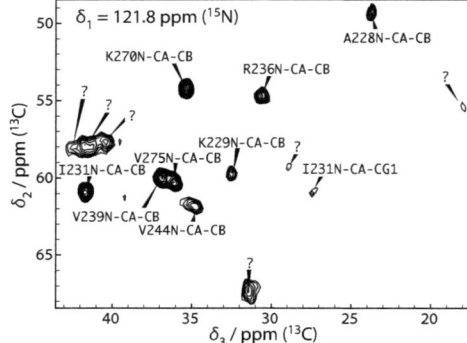

Figure II.3: Two-dimensional plane from the 3D NCACB correlation spectrum of HET-s. Peaks belonging to the core that are expected to appear for HET-s(218-289) are labeled with their respective assignments. The additional peaks are expected to arise from the N-terminal 217 residues and are labeled by question marks.

was recorded for HET-s. This spectrum (a representative 2D-plane is shown in Fig. II.3) asserts that the peaks of HET-s(218-289) (Wasmer et al. (2008a)) (BMRB ID 11028) are virtually all present in the spectra of HET-s fibrils, with the exception of six residues which also display only weak N–C^α or C^α–C^β peak intensities in the spectra of HET-s(218-289). A comprehensive list of assigned resonances is given in Table II.1. The measured chemical shifts in HET-s of nuclei belonging to the prion domain match those of HET-s(218-289) within 0.2 ppm for ^{13}C and 0.4 ppm for ^{15}N. Slightly higher deviations between the two samples for either backbone N or C^α resonances are marked in yellow and red, respectively, in Fig. II.4. Virtually all resolved signals from residues in the rigid core identified in HET-s(218-289) can be identified in HET-s, with the few exceptions shown in grey in Fig. II.4. Surprisingly, only 15 additional peaks not assignable to resonances from the prion domain show up in the 3D NCACB at the given signal-to-noise ratio, and all of them are weak compared to the peaks arising from the prion domain. This gross under-representation of the over 200 residues from the N-terminal domain must be due to static or dynamic disorder.

The comparison of the HET-s and the HET-s(218-289) spectra demonstrate clearly that the prion domain of an in-vitro refolded HET-s in its amyloid state has almost exactly the same structure as that of HET-s(218-289) in amyloid fibrils. The slight

Results 3

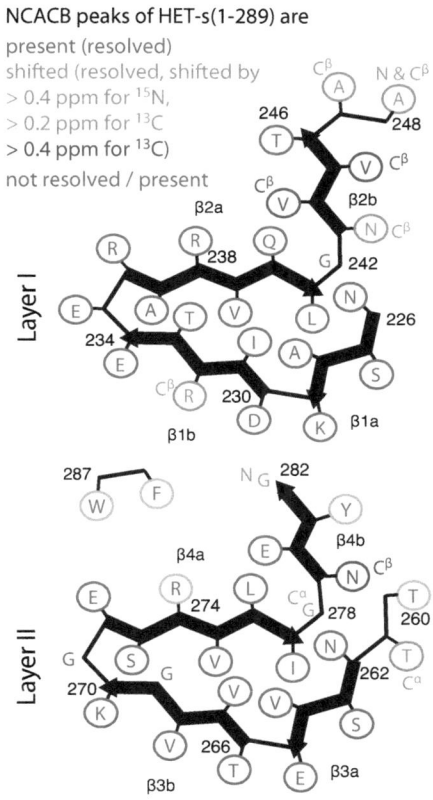

Figure II.4: Hydrophobic core of HET-s(218-289). The residue coloring indicates their appearance in the 3D NCACB spectrum (Fig. II.3) of HET-s(1-289).

53

variation in some chemical shifts may be due to minor differences between the two structures.

The structure of the globular domain is not conserved in the full-length HET-s fibrils. An overlay of extracts from 2D ^{13}C–^{13}C DARR (Takegoshi et al. (2001)) and ^{15}N–^{13}C NCA (Baldus and Meier (1996)) spectra of the HET-s fibrils (in black) with the spectra from the crystalline sample of HET-s(1-227) (in red) is given in Fig. II.1C-D and Fig. II.2B, respectively (full spectra are given in Fig. X.7). The spectra of the HET-s(1-227) sample reflect a well-ordered, well-structured, mainly α-helical protein that is in accordance with the X-ray crystal structure (Greenwald et al. (2010)). The HET-s spectrum (black) also features narrow resonances, most of which can be assigned to residues 218-289, as discussed above (see also Fig. II.1A, B, Fig. II.2A).

From the DARR and NCA spectra in Fig. II.1 and Fig. II.2 it can be seen that the

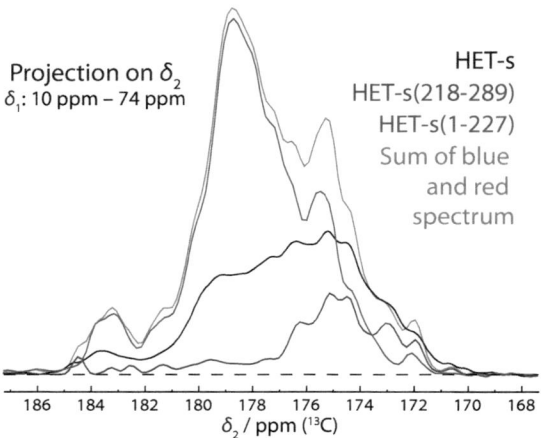

Figure II.5: Projections along the δ_1-dimension of the C$^\alpha$–C' region (δ_1: 10 ppm – 74 ppm, δ_2: 187 ppm – 168 ppm) of the DARR spectra (shown in Fig. II.1) of HET-s (black), HET-s(1-227) (red) and HET-s(218-289) (blue). Additionally, the sum of the HET-s(1-227) and HET-s(218-289) projections is given in pink. The spectra of HET-s(1-227) and HET-s(218-289) were scaled according to the number of observed residues (48 and 210, respectively). The HET-s spectrum (black) was scaled using the assigned peaks from the PD alone.

sum of the signals from the prion domain and HET-s(1-227) does not yield the resonance distribution displayed by full-length HET-s fibrils. Also, projections of the C^α–C' region (shown in Fig. II.5) reveal a resonance distribution for the full-length protein that is incompatible with the sum of the two parts and point to a decrease in α-helical structure with a relatively stronger contribution from residues with chemical shifts typical for backbone torsion angles in turn or β-sheet secondary structure for the N-terminal domain. Judging from the resolved resonances in the HET-s(1-227) spectra, the peak dispersion appears significantly reduced in HET-s (Fig. II.1C, D), indicating a loss of tertiary structure in the N-terminal domain of fibrillar HET-s compared to crystalline HET-s(1-227). The fibrillar nature of the sample has been confirmed by EM but so far, the different preparations have also shown a minor contamination by other aggregates (shown in Fig. S6 of Wasmer et al. (2009b)). These may or may not be fibrillar and could contribute to the signals that are attributed to the globular domain of HET-s. Pure aggregates can be pelleted by centrifugation before fibrils form and show only very weak signals in the CP spectra, and none at the HET-s(218-289) resonance position. Therefore the conclusions drawn here would not be influenced by a possible, but minor contamination.

Indications for a flexible linker connects the prion domain and the N-terminal domain in HET-s fibrils. It has previously been shown that HET-s(218-289) amyloid fibrils contain a flexible loop (Siemer et al. (2006b)) which can be detected in INEPT spectra (Lange and Meier (2008)). Similar ^{1}H–^{13}C INEPT experiments were performed on the HET-s, HET-s(218-289) and HET-s(1-227) samples (Figs. II.6 and II.7). The peaks were assigned to spin systems of individual residue types using experiments that employ, in addition to the initial INEPT from proton to carbon spins, a homonuclear ^{13}C–^{13}C TOBSY transfer step (Hardy et al. (2001)) (Fig. II.6). Peaks corresponding to most amino acids could be assigned (see below). All resonances are close to the average chemical shifts listed in the BioMagResBank (Ulrich et al. (1989)) and the random-coil chemical shifts (Wishart et al. (1995)) indicating that the mobility of these residues is due to a high local flexibility and not to large overall mobility of an entire domain.

The INEPT spectrum of crystalline HET-s(1-227) (given in Fig. II.7B) contains very few peaks, indicating that only the sidechains of some Lys and Thr residues are highly flexible. The finding that almost no flexible residues are detected is consistent with a well defined, rigid structure, and confirms that the globular domain assumes a quite different state when crystallized by itself or when in the

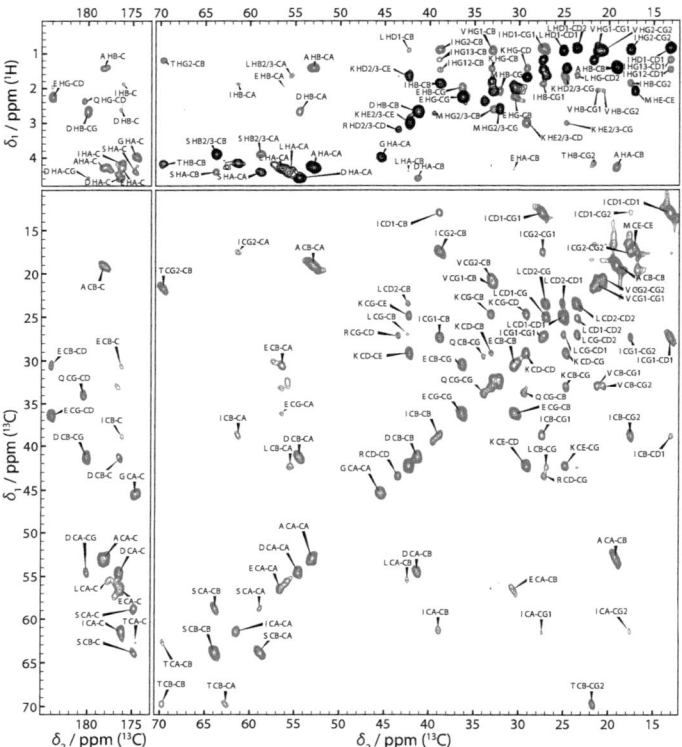

Figure II.6: Refocused, carbon-detected (Andronesi et al. (2005)) H–C INEPT (Burum and Ernst (1980)) spectrum (black) and INEPT-TOBSY (Hardy et al. (2001)) spectra (green) of HET-s. The two upper panels display the carbonyl and aliphatic regions of the HC INEPT and H(C)C INEPT-TOBSY experiment while the homonuclear (H)CC INEPT-TOBSY spectrum is given in the two lower panels. For the latter, an initial INEPT step is used to select for carbons in highly mobile regions which are then used as a starting point for a 2D homonuclear TOBSY correlation experiment.

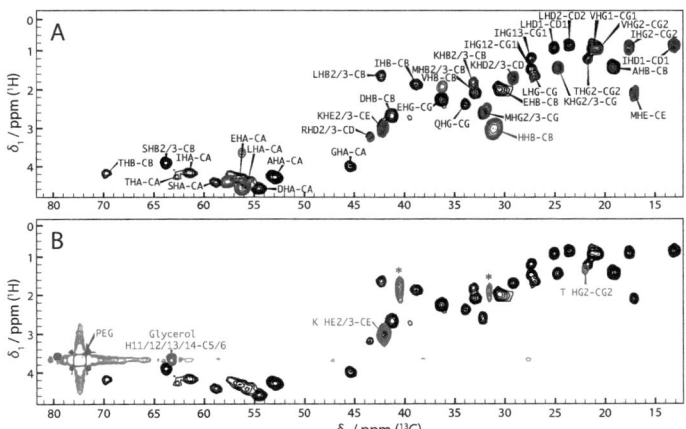

Figure II.7: **A** Refocused H–C INEPT spectra of HET-s(218-289) (blue) and HET-s (black). The spectrum of the HET-s fibrils features all peaks present for the (218-289) construct, except for the histidine $H^{\beta 2/3}$–C^β (labeled in blue) and two H^α–C^α cross-peaks. The black labels give the spin system assignments for the highly flexible parts of HET-s and were obtained by recording INEPT-TOBSY experiments (shown in Fig. II.6). Note that the M and V H^β–C^β peaks overlap here, but the resonances can be resolved in the INEPT-TOBSY experiments. An analysis of the peak intensities is given in Fig. II.8. **B** Refocused H–C INEPT spectra of HET-s(1-227) (red) and HET-s (black). The former features only 6 peaks of which 4 could be tentatively assigned based on their chemical shifts. Two peaks arise from polyethylenglycol (PEG) and glycerol contained in the sample and the two peaks marked by asterisks could not be assigned unambiguously.

context of the HET-s fibrils. To directly compare the INEPT spectra of HET-s and HET-s(218-289) we calibrated the spectral intensities using the N-terminal Met H^ϵ–C^ϵ cross peak. Because HET-s(218-289) contains only one Met residue but HET-s four, this calibration may lead to an under-, but not over-estimation of the relative HET-s peak intensities. Of all observed peaks, only those of Arg and Lys appear with comparable (less than a factor of 2 difference) peak intensities in the two spectra, while His is virtually only apparent for HET-s(218-289), which must be due to a loss in flexibility of the C-terminal His_6-tag in HET-s. On the other hand, Ala, Asp, Gln, Glu, Gly, Ile, Leu, Ser, Thr and Val appear only or with much higher intensity (at least by a factor of 5) in the HET-s spectra, indicating that the number of flexible residues of these types is significantly higher in the full-length protein. The relative peak volumes of all observed residues are plotted in Fig. II.8.

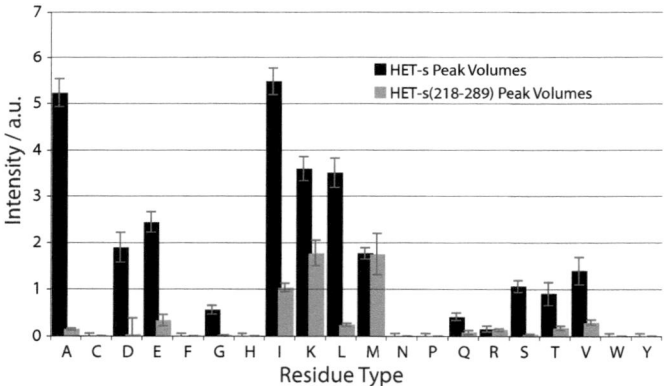

Figure II.8: Relative peak intensities (integrated peak volumes) in the INEPT spectra of HET-s(218-289) (gray) and HET-s (black). For amino acid types with more than 1 peak, the sum intensity is given. The Met H^ϵ–C^ϵ peak was set to equal intensity in both spectra in order to enable the comparison (see text). The intensity uncertainties, given by the error bars, were calculated based on the deviation of the Gaussian fit from the actual peak and the overall noise level of the spectrum. The uncertainty for the unobservable residues was estimated based on the noise level and the average linewidth.

The analysis of the INEPT spectra leads to some interesting conclusions. We note that no flexible Asn, Arg, Cys, His, Phe, Pro, Trp or Tyr residues could be detected (of Arg and His only the very end of the sidechain appears to be flexible) in the HET-s spectra, while the residue types Ala, Asp, Gln, Glu, Gly, Ile, Leu, Ser, Thr and Val are present with higher intensities than in the HET-s(218-289) INEPT spectrum. Because, with the exception of Thr_{260}, all prion domain signals are accounted for in the HET-s spectra that reveal the rigid parts, the residues giving rise to these enhanced signals must be located within residues 1-217 of the full-length HET-s fibrils. It can be deduced from the data that there is only one possible position for a single stretch of flexible residues that explains all observed signals: residues 191-217 would be comprised in this dynamically disordered region. Our data is also compatible with longer flexible stretches, ranging up to residues 181-224. In addition, the Val, Lys and Gln residues show no backbone resonances in the INEPT spectra of HET-s, indicating reduced flexibility. If we do not require those residue types to be flexible the arguments about the possible position of the flexible parts still hold. Most probably, additional stretches of flexible residues located within the N-terminal domain in the HET-s fibrils do also contribute to the INEPT spectra, consistent with the low overall intensity of resonances from this domain in the CP spectra and the presence of a molten-globule-like structure (see below).

4 Discussion

The NMR spectra of full-length HET-s in its amyloid form demonstrate that the prion domain, residues 218-289 of HET-s, forms an almost identical structural arrangement as in fibrils of the isolated prion domain. The fibrils investigated are thus organized around a highly ordered amyloid-like spine featuring a triangular hydrophobic core. While well-ordered amyloid cores have been predicted, not only for the HET-s protein, but also for Ure2p, Rnq1p and Sup35p (Wickner et al. (2008)), our data represents the first experimental observation of such a highly structured and well-ordered amyloid core for a full-length prion. The structural differences between the isolated prion domain and the domain in the context of fibrils of the full-length protein are very small, and they are predominantly located outside the region forming the triangular core. We thus conclude that the globular domain does not have much impact on the structure of the prion domain.

The globular domain, in contrast, undergoes considerable changes upon fibril formation. Crystals of HET-s(1-227) yield highly resolved NMR spectra indicating typical structural order of a crystalline protein and a predominantly α-helical fold. However, in the context of the full-length HET-s fibril, the N-terminal domain shows considerable structural disorder. The low signal intensity arising from the N-terminal domain in the CP-type spectra provide direct evidence for increased flexibility when compared to the C-terminal prion-domain. Despite this, overall the spectrum still indicates predominantly α-helical structural elements. Numerous residues in this domain are dynamically disordered in the context of HET-s fibrils and adopt a random-coil conformation. The evaluation of the INEPT spectrum is compatible with a stretch of 20-40 flexible residues centered around residue 200.

The reason for this partly disordered and dynamic state of the N-terminal domain might be found in the domain organization of HET-s in its soluble non-prion state and in the fibrillar prion state. While the former contains a well-defined globular domain comprising about residues 1-230 and a C-terminal tail that is highly dynamic (Balguerie et al. (2003)), the HET-s fibrils feature a highly ordered cross-β core region that includes residues 226-282. To accommodate this domain overlap, the globular domain needs to partially unfold upon fibril formation. This may indeed destabilize the entire domain: it has previously been reported that HET-s(1-217), i.e. HET-s minus the prion-domain, does not fold properly in vitro (Balguerie et al. (2003)), but adopts a molten globule-like state. This finding suggests that residues 217-227 are important for the folding and stability of the N-terminal domain. Our data supports the notion that in the context of HET-s fibrils, residues 1-217 contain native-like secondary structure elements, which however do not seem to be arranged into a rigid tertiary structure, as seen from the low dispersion of the signals. These characteristics infer that the N-terminal domain of the HET-s fibrils is a molten globule (Ohgushi and Wada (1983); Kuwajima (1989)).

Despite the large differences in their structures, the soluble and the fibrillar HET-s share some common features. In both states of the protein, a single, compact, well-defined folded domain is seen alongside with a second disordered part that lacks tertiary structure and is at least partially dynamic. Upon prion formation, the two domains switch roles, i.e. the previously dynamically disordered C-terminal part arranges into an highly ordered cross-β amyloid core while a loss of order is observed for significant parts of the globular domain. It can however be ruled out that the entire N-terminal domain is flexibly disordered and in a random coil conformation as it is the case for residues 229-289 in soluble HET-s (Balguerie et al. (2003)).

A structural model of the full-length fibrils, which is compatible with our results, is presented in Fig. II.9. We created the model starting from our previously calculated solid-state NMR structure of the HET-s(218-289) fibrils (Wasmer et al. (2008a)) (PDB entry 2RNM), composed of 10 HET-s molecule subunits. A flexible linker corresponding to the segment Lys218-Arg225 was chosen to connect each subunit of the β-solenoid fibrils (218-289) to the N-terminal part (1-217) of the X-ray structure of HET-s(1-227) (Greenwald et al. (2010)). To investigate the stability of the two-domain assembly, we performed an in vacuo simulation of the fibril composed of 10 full-length HET-s chains to obtain the displayed arrangement. We realized a short steepest descent energy minimization followed by a simulated annealing minimization with the AMBER96 force field (Cornell et al. (1996)) until convergence was reached, i.e. the energy improved by less than 0.1 % during 200 steps. There are no steric conflicts that would destabilize such a construct, and the model is compatible with the findings of this paper as well as the earlier H/D exchange data recorded for the prion domain in the fibrils of HET-s (Ritter et al. (2005)) and the observed dimensions in the electron-micrographs (Balguerie et al. (2003)). Therefore the proposed model in Fig. II.9 is energetically possible and sums up all experimental findings. To reflect the finding that the globular domain is still partly folded but has lost order, we represent it only by ellipsoids having the dimensions of the globular domain. From the Figure, it can immediately be seen that the globular domains are too large to be stacked on top of each other with the translational symmetry of the amyloid core, and that a reduction of the symmetry is required for the fibril formation of full-length HET-s.

Finally, it is interesting to note that the architecture of the HET-s amyloid fibrils is in contrast to the one of the yeast prion Ure2p, where the globular domain was found to be highly ordered in the fibrillar form, while the prion domain shows considerable disorder (Loquet et al. (2009)). The question arises whether order is possible for all domains in full-length prion fibrils, and if the order or disorder observed for the different domains is related to their amino-acid composition. One could also speculate if there might be an alternative form of the HET-s prion fibrils with the order/disorder content of the two domains exchanged.

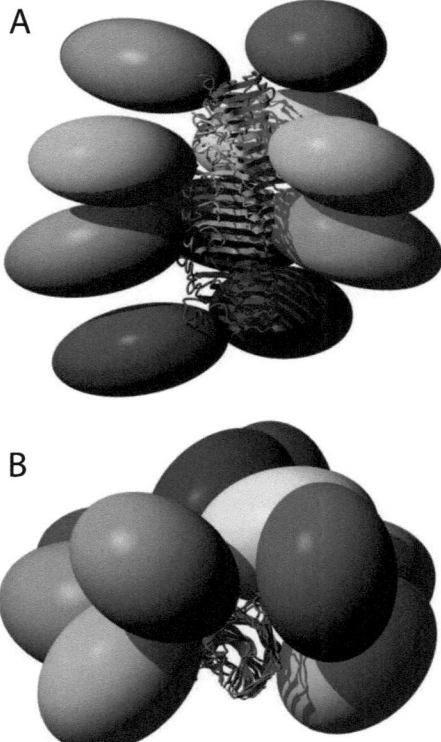

Figure II.9: Structural model of full-length HET-s. **A** Side view and **B** top view of 10 HET-s monomers within a HET-s amyloid fibril. The ellipsoids represent the N-terminal domains (residues 1-217) whose structure is not precisely known in this context. Each molecule is colored uniquely.

Chapter III

Non-Infectious pH 3 Fibrils of HET-s(218-289)

This work was done in collaboration with Alice Soragni and is published (Wasmer et al. (2008b)). Alice Soragni optimized the fibrillation conditions.

1 Introduction

Sabate et al. (2007) could recently show that the prion forming domain comprising residues 218-289 of HET-s forms amyloid fibrils at low pH *in vitro*, which have little or no infectivity. Because the structure of the infectious pH 7 form is known (Wasmer et al. (2008a)), the study of the non-infectious low-pH fibrils opens an exciting possibility to adress, on a molecular level, the differences that distinguish infectious from non-infectous polymorphs of the same protein.

Amyloids in general, and prions in particular, are known to exist in different polymorphic forms, partially controllable *in vitro* by external conditions like pH or stirring of the solution (Petkova et al. (2005); Paravastu et al. (2006)). Polymorphs can also be inheritable, a phenomenon which is intimately linked to the existence of different strains in prion diseases. Prion strains showing significantly different biological activity have been described in yeast by Baxa et al. (2006), Diaz-Avalos et al. (2005), and Toyama et al. (2007), but for the HET-s prion protein of the filamentous fungus *Podospora anserina* no indications for polymorphism under physiological pH conditions have been found (Sabate et al. (2007)). This finding is reflected in the solid-state NMR spectra of the prion-forming C-terminal domain of HET-s, the fragment HET-s(218-289), for which narrow NMR linewidths for both ^{13}C and ^{15}N resonances have been found, and no indications for peak doubling were detected (Siemer et al. (2005)). Furthermore, no variation of chemical shifts has been found over several preparations. For the pH 3 fibrils, whose NMR spectra are described in this communication, there is however evidence from EM, that there are indeed several coexisting polymorphs, all different from the pH 7 form (Sabate et al. (2007)).

The C-terminal fragment comprising residues 218 to 289 forms the proteinase K-resistant part of the fibrils (Balguerie et al. (2003)). As shown by Maddelein et al. (2002), it is necessary and sufficient for prion infectivitiy and forms infectious amyloid fibrils at pH 7 *in vitro* (Balguerie et al. (2003)). The well-resolved NMR spectra of the HET-s(218-289) pH 7 fibrils allowed for an almost complete sequence-specific NMR resonance assignment for the rigid parts (Siemer et al. (2006b)). The derived chemical shifts together with additional biophysical data have been used to propose a structural model with four beta strands, β_1 to β_4. These four strands form a β-solenoid fold with two repeating strand-turn-strand motifs (β_1-β_2 and β_3-β_4) forming two windings of the solenoid. The model is supported by recent electron microscopy data which show a mass-per-length ratio consistent with two layers of β-strands per HET-s(218-289) subunit (Sen et al. (2007)).

2 Materials and Methods

2.1 Sample Preparation

M-HET-s(218-289)-H_6 was recombinantly expressed and purified as previously described by Balguerie et al. (2003) and Siemer et al. (2006b). After purification the unfolded monomers were incubated at pH 3 in a buffer containing 40 mM boric acid, 10 mM citric acid, 6 mM NaCl at 37 °C over one week. The fibrils were pelleted by centrifugation (10,000 g), washed in pure water and centrifuged into the NMR rotor at 150,000 g. The pH 3 fibrils were found to be more stable throughout the experiments than fibrils formed at pH 2, where most of the experiments in (Sabate et al. (2007)). were performed. However, the pH 2 and pH 3 fibrils behave very similarly (both have almost identical aggregation kinetics, induce thioflavin-T fluorescence and show the same morphology in electron micrographs) and it was confirmed that the pH 3 form is not infectious (personal communication with S. Saupe).

2.2 Experimental details.

All spectra were recorded on a Bruker AV600 spectrometer operating at a static magnetic field of 14.09 T using 4 mm and 2.5 mm Varian T3 probes, processed in XwinNMR or TopSpin (Bruker Biospin) and analyzed using Sparky version 3.113 (T. D. Goddard and D. G. Kneller, University of California, San Francisco).

3 Solid-State NMR on pH 3 Fibrils

3.1 CP-based experiments

The ^{13}C–^{13}C proton-driven spin diffusion (PDSD) spectrum of HET-s(218-289) pH 3 fibrils is shown as blue contour lines in Fig. III.1. For comparison, the spectrum of pH 7 fibrils is given in red contours. After formation at pH 3, the pH 3 fibrils were washed in pure water. While fibrillization at neutral pH yields the pH 7 conformational state (Sabate et al. (2007)), the pH 3 form is stable at higher pH and

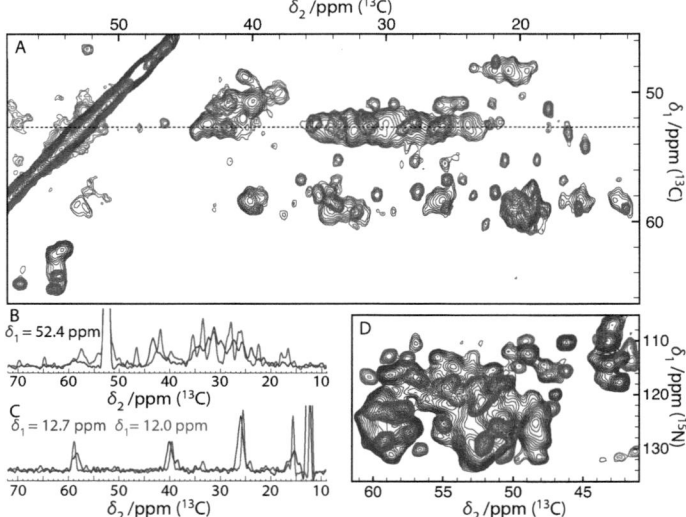

Figure III.1: PDSD spectra of HET-s(218-289) fibrils formed at pH 3 (blue) and of fibrils formed at pH 7 (red). **A** section of the aliphatic region (the complete spectra are shown in Figs. X.1 and X.2 in the appendix). **B, C** slices through a crowded part of the C^α-region and at an isoleucine $C^{\delta 1}$-resonance to clarify the differences in observed linewidths. **D** section of the ^{15}N–^{13}C HETCOR spectrum. All spectra were recorded with a mixing period of 50 ms and 90 kHz SPINAL-64 ^{1}H-decoupling during t_1 and t_2 at a static magnetic field of 14.09 T.

no pH 7 fibrils could be detected by our experiments. The spectra of the pH 3 fibrils are clearly different, indicating a different molecular structure, and the spectral resolution is somewhat lower, indicating higher disorder: The pH 3 fibrils exhibit typical linewidths between 128 Hz and 202 Hz compared to less than 100 Hz for the pH 7 fibrils (only well resolved peaks were analysed). The reduced resolution made a sequential resonance assignment difficult. Nevertheless, we have been able to identify and tentatively assign 22 spin systems, each corresponding to an amino acid residue, by through-bond ^{13}C–^{13}C TOBSY spectroscopy (Baldus and Meier (1996); Hardy et al. (2003)) (Fig. III.2), in combination with the PDSD and HETCOR spectra of Fig. III.1. The 22 spin systems detected in the rigid parts of the fibril consist of 3 A, 1 D (or N), 2 E, 2 G, 1 H, 2 I, 1 K, 1 L, 1 R, 2 S, 2 T and 4 V. The complete TOBSY spectrum and the assigned chemical shifts are given in the the appendix, in Fig. X.4 and Table X.1.

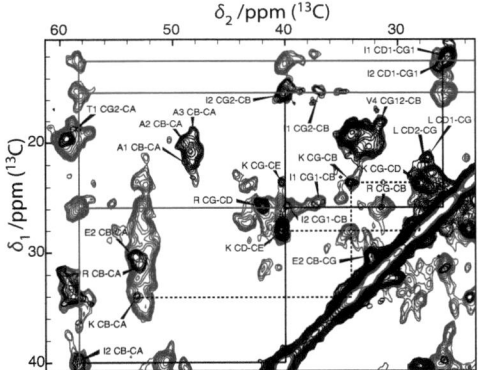

Figure III.2: Section of the ^{13}C–^{13}C-TOBSY (black) and 50 ms PDSD (blue) spectrum of the pH 3 fibrils. The continuous black and blue lines connect cross peaks belonging to I$_2$ in the TOBSY and the PDSD spectrum, respectively; the dotted black line follows cross peaks of the K spin system. The assignments were obtained by analysis of the TOBSY, PDSD and N–C HETCOR spectra and are numbered arbitrarily. The spectra shown were recorded at 13 kHz MAS with 90 kHz SPINAL-64 ^{1}H-decoupling during t_1 and t_2. The mixing times were 5 ms and 50 ms for the TOBSY and the PDSD spectrum, respectively.

For 16 of these spin systems, both C^α and C^β chemical shifts were assigned and their differences in secondary chemical shifts, $\Delta\delta_{C^\alpha} - \Delta\delta_{C^\beta}$, are shown in Fig. III.3.

For all of the spin systems negative values were found, indicating β-sheet structure (Spera and Bax (1991)) in accordance with FTIR data for pH 2 fibrils (Sabate et al. (2007)).

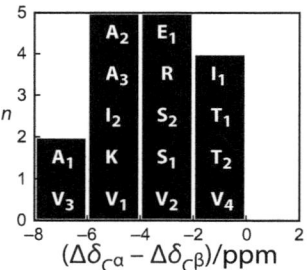

Figure III.3: Histogram of observed differences between C^α and C^β secondary chemical shifts. Negative values indicate β-sheet structure (Spera and Bax (1991)). The corresponding spin systems are given, using the same, arbitrary numbering as in Fig. III.2, X.4 and Table X.1.

3.2 INEPT-based experiments

To test for the presence of flexible residues in the pH 3 fibrils, ^{13}C-detected ^{1}H–^{13}C refocused INEPT and ^{1}H–^{13}C–^{13}C INEPT-TOBSY (Andronesi et al. (2005)) experiments were performed (Fig. III.4). Such experiments show only flexible parts of the fibrils (Siemer et al. (2006a)). Using the TOBSY connectivities, two Histidines and one Lysine side-chain were identified. In contrast to the pH 7 fibrils, no evidence for a flexible loop could be found (Siemer et al. (2006a)).

From the NMR experiments described above, we deduce that the HET-s(218-289) pH 3 fibrils consist of rigid β-sheets. In contrast to the infectious pH 7 fibrils, no highly flexible parts (except the HIS_6-tag) could be detected. There are a number of additional significant differences, indicating that the detailed structure of the pH 3 and pH 7 fibrils must be quite different. As seen in Fig. III.1 (and Figs. X.1 and X.2 in the appendix), the alanine C^α-C^β region, for example, consists in the pH 7 fibrils, of four strong peaks assigned to A228 (47.4, 21.6 ppm), A237 (51.1, 17.7), A247 (54.1, 14.8) and A248 (53.0, 16.1). Only A228 shows a shift inside the β-

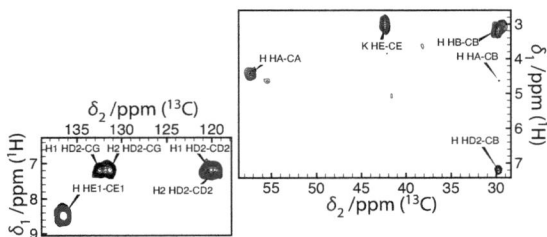

Figure III.4: Aliphatic and aromatic regions of the HC-INEPT (blue) and HCC-INEPT-TOBSY (black) spectra with tentative assignments based on the TOBSY cross-peaks and random-coil chemical shifts. Spin systems are numbered arbitrarily. Both spectra were recorded at 13 kHz MAS and 70 kHz SPINAL-64 ^{1}H-decoupling during t_2. The mixing time for the TOBSY spectrum was 4 ms.

sheet region. For the pH 3 fibrils, in contrast, no alanine resonance is detected outside the β-sheet region and we suspect that several alanines resonate in the partially resolved signal at (48.3, 21.3), indicating β-sheet. Further obvious differences appear in the serine C^α-C^β region and for valine (see e.g V264 C^α-C^β and C^α-C^γ, marked region in Figs. X.1 and X.2). Also, the overall quality of the CP/MAS spectra is different for the two types of fibrils. The increased linewidths of fibrils formed at pH 3 (Fig. III.1 b, c) suggest that they are not as well-ordered as the pH 7 fibrils of the same peptide. The mesoscopic structural variability as observed by electron microscopy (Sabate et al. (2007)) may be one source of disorder leading to a different set of signals for each polymorph. In addition, local disorder, which broadens the lines from individual polymorphic forms, could also play a role. Similar linewidths as for the pH 3 samples have been reported for other amyloids. (Petkova et al. (2002); Chan et al. (2005); Laws et al. (2001)) We conclude that the pH 3 non-prion amyloids of HET-s(218-289) have a rigid part almost exclusively found in β-sheet conformation but – in contrast to the infectious pH 7 form – does not contain flexible residues. Also, the structure of the individual HET-s(218-289) molecule embedded in the fibrils of the non-prion form appears to be quite different from the corresponding one in the prion form while the elementary fibril thickness and mass-per-length have been found to be similar. (Sabate et al. (2007); Sen et al. (2007)) Our results suggest that low pH fibrils lack prion-infectivity because their structure differs substantially from the one accessible under physiological pH conditions. Also consistent with this view is the fact that low pH fibrils are poor templates for HET-s *in vitro* fibrillization at pH 7 (Sabate et al. (2007)).

Chapter IV

HET-s(218-289) Inclusion Bodies

This work was done in collaboration with Laura Benkemoun, Raimon Sabaté, Michel O. Steinmetz, Bénédicte Coulary-Salin and Lei Wang and is published (Wasmer et al. (2009a)). Laura, Raimon and Bénédicte performed the biological experiments, Michel produced the electron micrographs and Lei recorded the H/D-exchange data.

1 Introduction

Heterologous protein expression in *E. coli* has taken a tremendous importance in many fields of biology and allows the preparation of sufficient amounts of sample for physico-chemical investigations (Baneyx (1999)). Often, however, expressed proteins are deposited as insoluble aggregates designated inclusion bodies (IB) (Baneyx and Mujacic (2004)) The formation of IBs, dense intracellular deposits of amorphous appearance that can reach $1\,\mu m^3$ in volume (Georgiou and Valax (1999)), is often regarded a mishap preventing the isolation of native heterologous proteins from *E. coli* cells. But IB formation can also be beneficial in some applications as it allows the recovery of high amounts of pure protein. In the context of structural biology, the deposition of aggregated protein in IBs is of interest because it may be closely related to the formation of amyloid deposits associated with many diseases (Carrio et al. (2005)). However, while IBs were often looked at as unspecific aggregates, fibrils were in contrast thought to be ordered. The notion now emerges that IB formation and fibrilization are both nucleation driven processes governed by sequence specific protein interactions which lead to the formation of specific β-sheet rich aggregates. Recent structural studies using quenched H/D exchange by NMR, fiber diffraction, microscopy, IR and mutagenesis have postulated that inclusion bodies are closely related to amyloids (Ventura and Villaverde (2006); Morell et al. (2008); Wang et al. (2008)). However, a high-resolution structural comparison of a protein in both states has not been available before this study. In prions the protein aggregation becomes self-perpetuating *in vivo* and is thought of as a fundamental process for infectivity. [Hets] is a prion of the filamentous fungus *Podospora anserina* involved in a fungal self/non-self recognition phenomenon (Coustou et al. (1997)). Previous studies have identified the C-terminal region of HET-s spanning residue 218 to 289 as the prion-forming domain (Balguerie et al. (2003); Ritter et al. (2005)). In its infectious amyloid form HET-s(218-289) forms a β-solenoid with two layers of β-strands per monomer, characterized by the formation of a triangular hydrophobic core (Wasmer et al. (2008a)). [Het-s] prion infectivity can be observed in in vitro assembled fibrils (Maddelein et al. (2002); Sabate et al. (2007)) and seems to be associated with the structurally characterized amyloid fibril type obtained at neutral pH. Alternate aggregated forms of HET-s including amyloid forms obtained at low pH lack [Het-s] prion infectivity (Sabate et al. (2007); Wasmer et al. (2008b)) in the sense of ref (Maddelein et al. (2002)). Herein, we describe a structural and functional characterization of HET-s(218-289) IBs produced in *E. coli* using electron microscopy, H/D exchange by NMR, solid-state NMR and in vivo prion infectivity assays as described earlier (Benkemoun et al. (2006)).

2 Materials and Methods

2.1 Inclusion body purification

DYT medium was inoculated at 37 °C with an overnight culture of Bl21(DE3) pLysS bacteria bearing the plasmid to be expressed (pET- 24a HET-s(218-289) or pET- hNDPk). When an OD_{600} of 0.5 was reached, the bacteria were induced with 1 mM of Isopropyl β-D-1-thiogalactopy-ranoside (IPTG) for at least 3 h at 37 °C. Then the culture was centrifuged and the cell pellet was frozen at –20 °C. Inclusion bodies were then partially purified under native conditions. After lysis, IBs were centrifuged twice for 30 min at 15000 g and washed in native buffer (50 mM Tris-HCl pH 8, 150 mM NaCl), then frozen at –80 °C. IB concentrations was estimated on an SDS-PAGE in comparison with HET-s(218-289) fibers of known concentration.

2.2 Sample preparation for EM

For the image shown in Fig. IV.1, *E. coli* pellets were placed on the surface of a copper EM grid (400 mesh) that had been coated with formvar. Each grid was very quickly submersed in liquid propane precooled and held at –180 °C by liquid nitrogen. The loops were then transferred in a pre-cooled solution of 4% osmium tetroxide in dry acetone in a 1.8 ml polypropylene vial at –82 °C for 72 h (substitution), warmed gradually to room temperature, followed by three washes in dry acetone. Specimens were stained for 1 h in 1% uranyl acetate in acetone at 4 °C. After another rinse in dry acetone, the samples were infiltred progressively with araldite (epoxy resin Fluka). Ultrathin sections were contrasted with lead citrate. Purified HET-s IBs (same sample as for solid-state NMR, see below) were an-

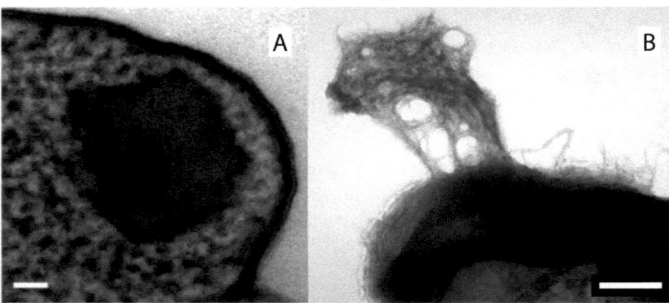

Figure IV.1: A HET-s(218-289) *E. coli* inclusion body observed by cryo EM of entire E.coli cells. **B** Transmission electron micrograph of negatively stained purified HET-s(218-289) IBs. In this electron micrograph evidence for fibrillar structures is found. Scale bars: 50 nm in **A** and 200 nm in **B**.

alyzed by EM four months after the expression in E. coli (the sample was stored at 4 °C during this period). The stability of the sample was checked by NMR which yielded identical spectra immediately after expression and after storage. For negative staining, used in Fig. IV.1A, sample aliquots of 5 µl were applied to a weakly glow-discharged carbon coated 200 mesh/inch copper grid. The sample was allowed to adsorb for 30 s, washed twice with water, and negatively stained for 20 s with 2% (w/v) uranyl acetate. Specimens were examined in a Philips Morgagni transmission electron microscope (TEM) operated at 80 kV. Micrographs were recorded with a Megaview III CCD camera at a nominal magnification of 50,000 ×.

2.3 Inclusion Body samples for solid-state NMR and H/D-exchange

Isotopically [^{13}C, ^{15}N] labeled M9 minimal medium was inoculated with a preculture of BL21(DE3) *E. coli* containing the pET-24a HETs(218-289) plasmid. At an OD_{600} of 0.6, the bacteria were induced with 1 mM IPTG for at least 5 h at 37 °C. The cells were collected by centrifugation and lysed using an M-110S Microfluidizer (Microfluidics Corp., Newton, MA, USA). The first sample ("raw IBs", Fig. 2 in the main text) was prepared by washing the cell extract 3 times in pure water. Further purification was achieved by subsequent washing in 8 M urea, 2% Triton X-100 and several times in water again to remove the Triton X-100. This procedure yielded the final sample (Fig. 3 in the main text). Both samples were centrifuged into 4 mm Chemagnetics thin wall rotors (sample volume 120 µl). The NMR spectra are shown in Figs. X.8 and X.9 in the appendix. To ensure that the inclusion bodies did not convert to amyloid fibrils throughout the expression period, we also prepared a sample by collecting the cells already 1 h after the induction by IPTG. These cells were as well lysed and washed (as the "raw IBs"). The NMR spectrum of this sample is shown in Fig. X.11.

2.4 Infectivity assays

The mechanical shearing method uses a cell disruptor (Fast-prep FP120, Bio101, Qbiogen Inc.). For each test, 0.5 cm^3 [Het-s*] mycelium is sheared (run time 30 s, speed 6 m/s) with 500 µl of STC50 buffer (0.8 M sorbitol, 50 mM CaCl$_2$, 100 mM Tris-HCl pH 7.5) and the sonicated protein or partially purified IB of interest (20 µl at indicated concentration) in a 2 ml screw cap tube. The sheared mycelium is then diluted with 600 µl of STC50 buffer, plated onto multiwell plates of DO-sorbitol 0.8 M and then incubated at 26 °C for 7-8 days. Then several implants (at least two per well) are checked for the [Het-s] phenotype in a barrage test against a [Het-S] mycelium as previously described.

2.5 Solid-state NMR spectroscopy

All spectra were recorded on a Bruker Avance spectrometer with 600 MHz proton frequency (B_0 = 14.9 T) equipped with a 4 mm Chemagnetics T3 probe. The MAS frequency was stabilized at 10 kHz, the probe temperature was –5 °C (sample temperature ca. +3 °C). The ^{13}C–^{13}C correlation spectra (Figs. X.8, X.9, X.10 and X.11) were obtained using a 50 ms PDSD mixing period, and SPINAL-64 ^{1}H-decoupling at 100 kHz during t_1 and t_2. The maximum t_1 and t_2 Evolution times were 15 ms, the spectral width was 50 kHz in both dimensions. The CP step used contact times of 400 µs (IB sample) and 800 µs (raw IBs and *in vitro* fibrils), ^{1}H RF-fields of 96 kHz, 46 kHz and 58 kHz and ^{13}C RF-fields of 100 kHz, 64 kHz and 62 kHz for the IB sample, the raw IBs and *in vitro* fibrils, respectively. The recycle delay was 2.5 s and 32 scans were performed leading to a total experiment time of 35 h.

2.6 H/D-exchange by liquid-state NMR

HET-s(218-289) *in vitro* amyloid fibrils and purified inclusion bodies (same sample as for solid-state NMR, see above) were H/D exchanged for 13 h at 4 °C before measurement. The H/D exchange buffer contained 150 mM Nacl, 50 mM Tris pH 8.0, 99.6% D_2O. Fast HMQC spectra of the fibrils and inclusion bodies after the exchange period in d_6-Dimethyl sulfoxide (DMSO) containing 0.1% d_1-Trifluoroacetic acid (TFA) are shown in Fig. X.12. For comparison, a spectrum was also taken before the exchange period. The result is virtually identical for in vitro fibrils and IBs.

3 Bio-Chemical Experiments

HET-s(218-289) was expressed in *E. coli* cells which show typical electron-dense, angular-shaped inclusions, 100 nm to 400 nm in cross section, located at cellular poles (Fig. IV.1). IBs were partially purified (see Appendix) and this material was analyzed by EM (Fig. IV.1B). Fibrillar structures can be clearly seen in this preparation. The purified material was then used to infect prion-free *Podospora anserina* strains using a protein transfection method as previously described (Benkemoun et al. (2006)). While insoluble extracts from an *E. coli* control strain containing only the empty vector or IBs of a different protein (human NDP kinase) did not induce appearance of the prion form [Het-s] (Table IV.1), strains transfected with HET-s(218-289) IBs acquired [Het-s] at a frequency comparable or even higher than that for transfection with HET-s(218-289) amyloids assembled in vitro. We conclude from these experiments that HET-s(218-289) IBs display [Het-s] prion infectivity, indicating that HET-s(218-289) acquires an infectious prion fold in *E. coli* IBs.

protein	amount of protein in transfection reaction /nmol					
	2.0	1.0	0.5	0.2	0.1	0.05
insoluble *E. coli* extract of strain with empty vector			3/138*			
NDP kinase IBs			5/84			
HET-s(218-289) fibrils		17/24	44/60	31/72	7/24	5/24
HET-s(218-289) IBs		23/24	58/60	34/72	15/24	13/24
HET-s(218-289) IBs (purified NMR sample)	24/24		45/48			

*The amount of E. coli insoluble extract used corresponds to the amount of cells that yield about 0.5 nmol HET-s(218-289) IBs.

Table IV.1: Prion infectivity assays with *E. coli* HET-s(218-289) IBs. For each case, the number of prion infected strains over the total number of tested strains is given.

Next, a set of experiments comparing further properties of HET-s IBs and amyloid fibrils assembled in vitro was performed. We found in particular that full-length HET-s or HET-s(157-289) inclusion bodies display the same 8 kDa proteinase K resistant core as in vitro assembled HET-s and HET-s(157-289) amyloid fibrils (Balguerie et al. (2003, 2004)). Moreover, both types of HET-s(218-289) aggregates, IBs and fibrils, display similar chemical denaturation properties. In particular, both are resistant to urea denaturation at neutral pH but highly sensitive to urea denaturation at pH 2 (supporting figure S2B in (Wasmer et al. (2009a))). Finally, we found that HET-s(218-289) IBs could act as seeds for in vitro amyloid formation of HET-s(218-289) (supporting figure S3 in (Wasmer et al. (2009a))) similar to HET-s(218-289) prion amyloids (Sabate et al. (2007)). It is also of note here that mutations in the 218-289 region that prevent the formation of the amyloid prion fold (Ritter et al. (2005); Maddelein (2007)) also prevent IB formation in *E. coli* (supporting table S1 in (Wasmer et al. (2009a))). Collectively, these results further emphasize the similarities between HET-s(218-289) IBs and HET-s(218-289) prion amyloid fibrils.

4 Solid-State NMR on HET-s(218-289) Inclusion Bodies

In order to characterize the HET-s(218-289) IBs on a molecular level, we recorded solid state NMR spectra and compared them to previously assigned spectra of the prion fibril assembled in vitro (Siemer et al. (2005)) whose structure has been determined (Wasmer et al. (2008a)). The so-called raw IBs, obtained by washing the insoluble fraction of the lysed cells in pure water, give rise to the ^{13}C-^{13}C proton-driven spin diffusion (PDSD) (Szeverenyi et al. (1982); Grommek et al. (2006)) spectrum shown in Fig. IV.2 (blue contours). The spectrum reproduces all peaks visible for HET-s(218-289) fibrils assembled in vitro (red contours). The separate spectra of the two samples are given in the Appendix, in Figs. X.8 and X.10, respectively. Additional resonances, not belonging to HET-s(218-289), appear in several regions of the spectrum and are tentatively assigned to phospholipids from the *E. coli* membrane (around 34 ppm the CH_2 groups of the hydrophobic tails and between 60 and 100 ppm (Fig. IV.2)) as well as to other proteins and possibly RNA. The many carbonyl resonances with chemical shifts in the range of 176 ppm to 180 ppm, where only a few HET-s(218-289) resonances are present, indicate the abundance of proteins with mainly α-helical secondary structure (Wang and

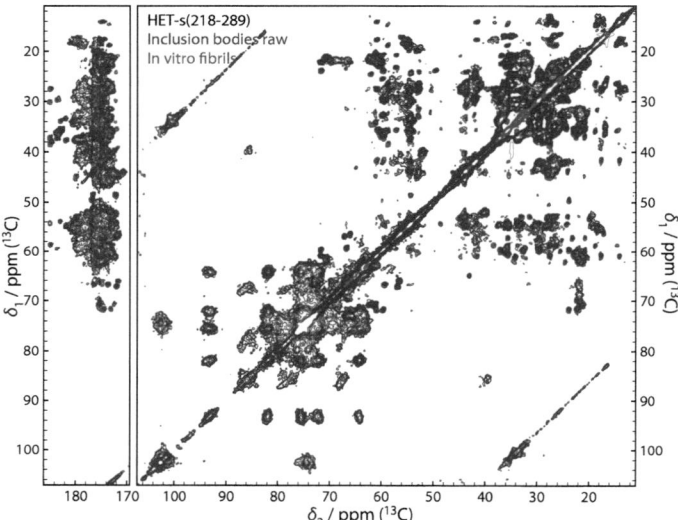

Figure IV.2: ^{13}C-^{13}C solid-state-NMR correlation spectrum (PDSD with a mixing time of 50 ms) of raw HET-s(218-289) IBs (blue contours) compared with a spectrum under identical conditions for purified and in vitro fibrillized HET-s(218-289). All signals assigned for the purified fibrils were also observed in the IB Spectrum. The separate spectra are shown in the Appendix, Figs. X.8 and X.10.

Jardetzky (2002)), most probably membrane proteins. Note that the peaks assigned to HET-s(218-289) are already visible after a short expression time of 1 h (see Appendix, Fig. X.11). After further purification of the sample (see Appendix) all HET-s(218-289) resonances in the PDSD spectrum are preserved while the additional components are considerably reduced (Fig. IV.3). The α-helical proteins seem to be almost completely removed by the further purification procedure (see carbonyl region). Considering the strong dependence of the NMR chemical shift on the conformation of a polypeptide chain, the identical shifts of HET-s(218-289) in IBs and in vitro fibrils show that their molecular structures have to be virtually identical. Noticeably, the NMR linewidth of in vitro fibrils and IBs is also indistinguishable, as judged from isolated signals in the 2D spectra (see Fig. IV.4 and insets of Fig. IV.3), indicating a highly defined local molecular structure (Siemer et al. (2005)). No indications for molecular polymorphism were found. The purified IB sample prepared for NMR was also found to display prion infectivity in transfection assays (Table IV.1). In addition to the solid-state NMR characterization, an H/D-exchange experiment was performed as previously used for the characterizion of the amyloid fibrils of HET-s(218-289) (Ritter et al. (2005); Hoshino et al. (2002)). The observed exchange pattern of the purified inclusion bodies closely resembles that of the in vitro fibrils of HET-s(218-289) (Appendix, Fig. X.12), which further supports and verifies, that these are virtually identical on a molecular level.

5 Conclusions

Our results indicate that the highly ordered amyloid HET-s(218-289) can be formed in the crowded milieu of the *E. coli* cell in the presence of folding modulators, chaperones and high amounts of other proteins. This observation is of particular importance in the context of prion structural biology as it indicates that the fold characterized so far only in vitro, can also form in a living cell. This also shows that inclusion bodies can be highly ordered in contrast to being "amorphous" as could be inferred from their appearance in electron micrographs of in vivo material. The solid-state NMR spectra show that HET-s(218-289) adopts the same molecular structure in IBs and in vitro fibrils and that the IBs therefore consist in fact of amyloids. A careful inspection of electron micrographs gave indeed clear evidence for fibrillar structures. Whether HET-s(218-289) represents an exception

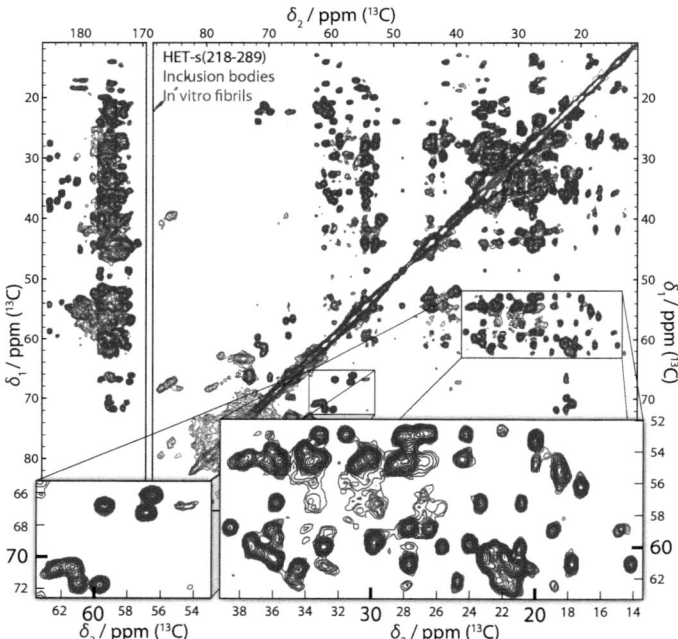

Figure IV.3: ^{13}C-^{13}C solid-state-NMR correlation spectrum (PDSD with a mixing time of 50 ms) of purified HET-s(218-289) IBs (blue contours) compared with a spectrum under identical conditions for purified and in vitro fibrillized HET-s(218-289). All signals assigned for the purified fibrils were also observed in the IB Spectrum. The insets and the 1D (Fig. IV.4) demonstrate that the linewidth of the two samples is virtually identical and that no significant chemical-shift changes appear. The separate spectra are shown in the Appendix, Figs. X.9 and X.10.

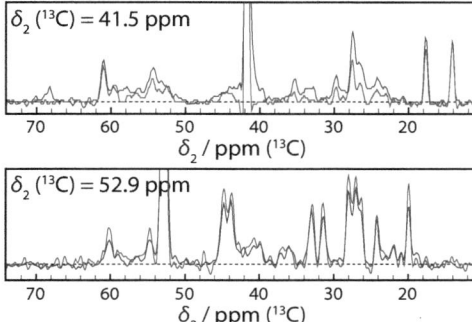

Figure IV.4: Slices form the PDSD Spectra (Fig. IV.3) of purified HET-s(218-289) IBs (blue contours) and in vitro fibrillized HET-s(218-289). The linewidths in the two spectra are virtually identical.

or whether other proteins can also display a highly ordered and functional amyloid structure in the bacterial inclusion body remains to be determined. For HET-s, in any case, it is obvious that the formation of IBs and amyloid fibrils can be a remarkably similar process and that the inclusion bodies are infectious and highly ordered at the molecular level. Evidence for amyloid-like behaviour was already found for IBs of the Alzheimer peptide Aβ(1-42) (Morell et al. (2008)) and other proteins (Wang et al. (2008)) and our detailed atomic-level investigation for the HET-s system supports the emerging concept that amyloid formation is ubiquitous in living organisms and must be considered in biotechnological protein production.

IV HET-s(218-289) Inclusion Bodies

Chapter V

Structural Similarity between the Prion Domain of HET-s and a Homologue—FgHET-s

This work was done in collaboration with Agnes Zimmer, Raimon Sabaté and Alice Soragni and is published (Wasmer et al. (2010)). Agnes collected the H/D-exchange data, Raimon performed the biochemical experiments and Alice performed additional cross-seeding experiments.

1 Introduction

Prions are infectious particles composed solely of protein (Prusiner et al. (1998)). In addition to the disease causing mammalian prions, prions have also been identified in yeast and fungi. These prions represent interesting model systems to study the process of prion propagation (Wickner et al. (2007)). The [Het-s] prion of the filamentous fungus Podospora anserina is involved in a non-self recognition process termed heterokaryon incompatibility that operates when strains of unlike genotypes fuse and which leads to cell death of the fusion cell (Coustou et al. (1997)). The het-s gene locus has two alternate incompatible alleles designated het-s and het-S that encode for the proteins HET-s and HET-S, respectively. Strains expressing HET-s in its soluble form are termed [Het-s*]; strains expressing the fibrillar prion form of HET-s are designated [Het-s]. It is the prion form [Het-s] that shows the heterokaryon incompatibility reaction with [Het-S].

HET-s represents an attractive model to study the sequence-structure relationship in amyloidal prions. Fibrils formed *in vitro* from the prion domain HET-s(218-289) (Balguerie et al. (2003)) feature a highly-ordered, triangular amyloid core of which an atomic-resolution structure has been determined (Wasmer et al. (2008a)). It can be described as a β-solenoid (see ref. (Kajava and Steven (2006)) for definition) where one molecule forms two windings. In addition to the rigid, highly ordered core region, HET-s(218-289) also contains a dynamically disordered flexible loop, comprising residues 250-259 (Siemer et al. (2006a); Lange et al. (2009)). This fold of the isolated prion domain is maintained in the context of the full-length prion (Wasmer et al. (2009b)).

In this manuscript, we describe a distant homologue of the fungal prion HET-s found in the filamentous euascomycete Fusarium *graminearum*, which is a prominent wheat, barley, oat and maize pathogen (Parry et al. (1995)). As HET-s, the homologue, which we denote by FgHET-s, comprises 289 amino acid residues but both proteins display a sequence identity of only about 50 % for all residues and 38 % for the prion domain (residues 218-289). While FgHET-s has not been tested for prion activity in its native host, we show below that recombinantly produced FgHET-s(218-289) can form amyloid fibrils *in vitro*. These fibrils are able to efficiently cross-seed HET-s(218-289) fibril formation (and *vice versa*). In the following, H/D-exchange and solid-state NMR data from FgHET-s(218-289) are found to be remarkably similar to those of HET-s(218-289) fibrils, despite the rather low sequence identity. Based on these data, we propose a structural model based on HET-s, which shares important features, like the hydrophobic core and lattices of water-exposed salt-bridges. Our findings provide a structural basis for the observed efficient cross seeding of the amyloid form.

2 Materials and Methods

2.1 Plasmids and strains

The *Fusarium graminearum het-s* homologue has been cloned by PCR on genomic DNA of strain PH-1 (NRRL 31084) (genomic DNA prep was a generous gift of Jin-Ron Xu, Purdue University) using the following primers; 5'TTCCAACAATAGC-TAACCGC3' and 5'ATTCAACACAGCCAACCGGC3'. The PCR fragment was cloned in the pGEM-T vector (Promega). The pET-24a-FgHET-s(218-289) plasmid was constructed by amplifying the region encoding for the C-terminal part of the protein (residue 218 to 289) by PCR using primers, 5'ATCATATGAAGTTGA-ACATGATCGAGG 3' and 5'ATAAGCTTAATGGTGATGGTGATGGTGATCTT-CCCAGATGCCTCTGCC3'. The PCR fragment was restricted by both NdeI and HindIII and cloned into the pET-24a vector (Novagen).

2.2 Protein expression

HET-s (218–289) (Dos Reis et al. (2002)) and FgHET-s(218-289) were expressed in 21 DYT medium at 37 °C by BL21(DE3) pLysS cells bearing the apporpriate plasmids. When an OD_{600} (optical density at 600 nm) of 0.6–0.8 was reached, the bacteria were induced with 1 mM isopropyl-1-thio-β-D-galactopyranoside (IPTG). After 3 h at 37 °C the cultures were centrifuged and the cell pellets were frozen at −20 °C.

2.3 Protein purification

HET-s(218-289) and FgHET-s(218-289), expressed with a C-terminal his-tag, were purified at denaturing conditions (50 mM Tris/HCl [pH 8], 300 mM NaCl and 6 M GuHCl buffer) by Ni-affinity chromatography on Talon his-tag resin (ClonTech).

The buffer was exchanged by gel filtration on Sephadex G-25 column (Amersham) for 175 mM acetic acid [pH 2.5] and the proteins were conserved at 4 °C.

Amylin peptide (QRLANFLVHSSNNFGAILSS) was obtained from EZ Biolab Inc. (Carmel, IN, USA). A 5 mM stock solution was prepared in 1,1,1,3,3,3-hexafluoro-2-propanol (HFIP), which had been sonicated two times for 30 min and dried at 4 °C and then had been centrifuged at $15,000g$ for 15 min and was finally filtrated by millex-GV 0.22 µm filters in order to remove possible residual quantities of large aggregates. After drying, the solution was incubated at room temperature for 10 min. Stock solutions were divided into aliquots (20 µl per eppendorf) and HFIP was removed by evaporation under a gentle stream of nitrogen, leaving a slight film; finally, the samples were stored at −80 °C. When required, the samples were resuspended in 50 µl of anhydrous dimethyl sulfoxide (DMSO) and were sonicated for 10 min. Sonication was crucial to remove any traces of non-dissolved seeds that may resist solubilization. This preparation yielded amylin in monomeric form. Aliquots of amylines were added to 100 µM acetate buffer [pH 5.5] and 850 µM of miliQ water obtaining a final peptide concentration of 100 µM. Peptide aggregation from soluble monomer was monitored by measuring the transition from non-aggregated to aggregated state by relative ThT fluorescence at 480 nm when exciting at 445 nm. Amylin aggregation was carried out at 37 °C with a soluble monomer concentration of 15 µM.

2.4 ThT-binding determination

Thioflavin-T (ThT) binding with HET-s(218-289) or FgHET-s(218-289) was measured using a Perkin-Elmer LS50 fluorescence spectrometer with an excitation wavelength of 450 nm and emission range from 470 nm to 570 nm and the emission at 480 nm was recorded. ThT and protein concentrationd of 25 µM and 10 µM, respectively, at pH 7 and 37 °C were used.

2.5 Electron microscopy

For electron microscopy, 400-mesh copper electron microscopy grids coated with a plastic film (Formvar) were used. A fraction of the protein suspension (at 1 mg/ml) was put onto the grid and sedimented during 10 to 30 min in a moist Petri dish to avoid rapid desiccation. Grids were then rinsed with 15-20 drops of freshly prepared 2 % uranyl acetate in water and filtered with 0.22 µm Millipore, dried with filter paper and observed with a Phillips TECNAI 12 Biowin electron microscope at 80 kV.

2.6 Aggregation assays

HET-s(218-289) and FgHET-s(218-289) aggregation from soluble monomers was monitored by concurrently observing the transition from the soluble to the aggregated state by UV/Vis absorbance at 280 nm (tryptophan-tyrosine peak plus scattering) and 400 nm (scattering of the sample). All experiments were carried out with 10 μM of soluble monomer at 25 °C and agitation every 5 min (by brief vortex pulse) in order to homogenize the samples. HET-s(218-289) fibrillations were realized at pH 7 (in a 1:1 mixture of 175 mM acetic acid and 1 M Tris/HCl pH 8) (Sabate et al. (2007)). *Fusarium* fibrillations were realized at pH 4 (in a 3:1 mixture of 175 mM acetic acid and 1 M Tris/HCl pH 8), in order to avoid the spontaneous aggregation of FgHET-s(218-289). For seeding and cross-seeding aggregation assays, 1 μM (representing 10 % of total protein concentration) of the respective other, preformed fibrils were added to an initially 10 μM protein solution. In addition, in order to confirm the seeding and cross-seeding capacity, 0.1 μM (1 % of total protein concentration) of HET-s(218-289) and FgHET-s(218-289) fibrils were tested.

2.7 Chemical denaturation curves

FgHET-s(218-289) and HET-s(218-289) stabilities in the presence of guanidine hydrochloride and urea were studied at pH 7. The fraction of denatured protein (f_D) was calculated from the fitted values using the equation: $f_D = 1 - ((y_D - y)/(y_D - y_N))$, where y_D and y_N are the fluorescence maximum wavelengths or the relative fluorescence (RF) at a fixed wavelength of the denatured and native protein, respectively, and y is the fluorescence maximum wavelength or RF at a fixed wavelength of protein as a function of denaturant concentration. A non-linear least-squares analysis was used to fit the denaturation curves to

$$y = \frac{(y_N + m_N \cdot [D]) + (y_D + m_D \cdot [D]) \cdot e^{A([D]-m_{1/2})/RT}}{1 + e^{A([D]-m_{1/2})/RT}},$$

where y represents the observed fluorescence maximum wavelength or RF at a fixed wavelength, y_N and y_D are the intercepts, and m_N and m_D are the slopes of the pre- and post-transition baselines, [D] is the chemical denaturant concentration, $m_{1/2}$ is the denaturant concentration at the midpoint of the curve, and A is a constant generated by the fitting (Santoro and Bolen (1988); Pace et al. (1998); Koditz et al. (2004)).

2.8 Hydrogen-Deuterium Exchange

U-[^{13}C, ^{15}N] and U-^{15}N labelled FgHET-s(218-289) were recombinantly expressed in *E. coli* and amyloid fibrils were prepared as described for HET-s(218-289) (Ritter et al. (2005)). ^{15}N-labelled FgHET-s(218-289) fibrils were used for H/D-exchange studies relating to the backbone amides (Hoshino et al. (2002); Li and Woodward (1999)). To start the exchange reaction, fibrils were pelleted at 20,800 g for 4 min, washed in 50 mM Tris·HCl pH 7.3 comprising 150 mM NaCl and D$_2$O as the solvent, pelleted again and re-suspended in the same buffer for incubation up to 12 weeks. Hydrogen exchange was quenched at suitable intervals by pelleting the fibrils at 20,800 g for 4 min and freezing the pellet on liquid nitrogen. For NMR-analysis the fibrils were solubilized in perdeuterated DMSO (d$_6$-DMSO) containing 0.05 % deuterated trifluoric acid (d$_1$-TFA). Afterwards a series of 80 2D [^{15}N, ^{1}H] correlation spectra was recorded for 4 h (3 min per spectrum) on a Bruker AVANCE III 600 spectrometer equipped with a CryoProbe unit. The amount of residual D$_2$O in DMSO was about 4 %. Residues displaying fast exchange in the fibrils as well as residues with high intrinsic exchange rates in DMSO result in absent peaks in the [^{15}N, ^{1}H] correlation spectrum. To identify the latter, a second series of 80 2D spectra were measured after the addition of 4 % H$_2$O. Using identical solvent conditions as for hydrogen-exchange NMR analysis, we carried out triple-resonance HNCACB (Wittekind and Mueller (1993)) and HNH NOESY (Diercks et al. (1999)) experiments on U-[^{13}C, ^{15}N] FgHET-s(218-289) to achieve the sequence-specific resonance assignment of the backbone amide moieties. All residues except for R252 could be assigned. [^{15}N, ^{1}H] correlation spectra (Fig. X.13) were used to quantify the residual protonation depending on incubation time by integrating the peak volumes. To determine the specific exchange rates, we fitted these data to a mono-exponential decay. The data were analyzed by using the programs PROSA (Guntert et al. (1992)) and CARA (Keller (2004)), and a specially written Visual basic program. The resonances of N243 and N279 overlapped strongly. Since both residues displayed fast exchange, and are located at identical positions within the repeat units, an average exchange rate for both residues was calculated.

2.9 Solid-state NMR

U-[^{13}C, ^{15}N] FgHET-s(218-289) was recombinantly expressed in *E. coli* and amyloid fibrils were prepared as described for HET-s(218-289) (Ritter et al. (2005)).

These were washed in water and centrifuged into a 3.2 mm NMR rotor at 200,000 g (Böckmann et al. (2009)). All solid-state NMR experiments were carried out on a Bruker AVANCE II+ wide-bore spectrometer with 850 MHz proton frequency ($B_0 = 20.0$ T) equipped with a 3.2 mm triple-resonance MAS probe. The MAS frequency was stabilized at 19.00 kHz, the sample temperature was ca. 3 °C and SPINAL64 proton decoupling of 100 kHz was applied for all spectra. 2D and 3D NCACX and 2D and 3D NCOCX spectra (Detken et al. (2001); Siemer et al. (2006b)) and a 2D C–C homonuclear correlation spectrum with 100 ms DARR / MIRROR mixing (Takegoshi et al. (2001); Scholz et al. (2008)) (simply called DARR in the rest of the text) were recorded. Each of the two 3D experiments was acquired within 4 days of measurement time, the 2D N(CA)CX and N(CO)CX within 3 days each, and the DARR spectrum in 14 h. The ^{13}C–^{13}C polarization transfer in between carbonyl and aliphatic carbons was found to be optimal for a ^{1}H RF field irradiation of about 15 kHz during the mixing period (neither exactly at the DARR nor the MIRROR condition). A length of 50 ms was chosen for these homonuclear transfer steps.

While the 2D spectra exhibit a higher signal-to-noise ratio, the resolution of peaks is superior in the 3D experiments. Most of the sequence-specific assignments were made using the NCOCX and NCACX 3D-correlation spectra while the ^{13}C–^{13}C DARR spectrum was primarily used for verification and side-chain assignments.

In order to detect highly flexible residues, experiments employing an initial IN-EPT (Morris and Freeman (1979); Burum and Ernst (1980)) with detection on ^{13}C (Andronesi et al. (2005)) were carried out at 17 kHz MAS, a sample temperature of 20 °C and SPINAL64 proton decoupling of 50 kHz at a static magnetic field of 20.0 T. In order to accomplish the spin-system resonance assignment, we added a ^{13}C–^{13}C transfer step to the initial ^{1}H–^{13}C INEPT. This homonuclear transfer was realized by a 5 ms TOBSY mixing period employing the $P9_3^1$ sequence (Hardy et al. (2001)) with an RF field of 102 kHz on ^{13}C. Topspin 2.0 (Bruker Bio- spin) was used to process all spectra and Sparky 3.113 (T. D. Goddard and D. G. Kneller, University of California, San Francisco) for the sequence-specific resonance assignment.

3 Results

3.1 Sequences homologueous to the HET-s prion domain exist in various Fusarium species.

Searching the available fungal genomic databases at NCBI and Broad Fungal Genome Initiative with the HET-s prion domain as query in BLASTP searches, homologueous sequences were identified in various Fusarium species, namely F. graminearum (*Gibberella zeae*), F. *verticilliodes* (*Gibberella moniliformis*), F. *oxysporum* and *Nectria haematococca* (Fusarium *solani*). An alignment of the sequences of the C-terminal region of *Fusarium* proteins showing homology to HET-s(218-289) is given in Fig. V.1. The closest homologue is found in F. *graminearum*. The predicted protein FG10600 was considered as the F. *graminearum* HET-s based on the reciprocal best hit method11 and will be referred to as FgHET-s in the following. Overall HET-s and FgHET-s show 50 % identity (55 % in the globular domain and 38 % in the region corresponding to the prion domain). The C-terminal region of FgHET-s (FgHET-s(218-289)) is the closest homologue to the HET-s prion domain identified in this search and was chosen for further characterization.

3.2 Recombinant FgHET-s(218-289) forms amyloid fibrils *in vitro*.

In order to analyze the properties of the FgHET-s prion domain, the region corresponding to the HET-s prion domain FgHET-s(218-289) was expressed as previously described for HET-s(218-289) with a C-terminal His_6 tag and purified under denaturing conditions from inclusion bodies (Balguerie et al. (2003)). Like HET-s(218-289), FgHET-s(218-289) remained soluble at pH 2.5 (175 mM acetic acid) but spontaneously aggregated into amyloid fibrils at pH 7 at 20 µM. Similar to HET-s(218-289) fibrils formed under the very same conditions, FgHET-s(218-289) forms bundles of laterally associated individual fibrils with a width of about 5 nm (Fig. V.2A). In contrast to HET-s(218-289) fibrils, which were reported not to induce ThT fluorescence by Sabate et al. (2007), FgHET-s(218-289) fibrils do induce a robust ThT fluorescence (Fig. V.2B).

The stability of FgHET-s(218-289) fibrils against denaturation by both GuHCl and urea was probed by measuring Tryptophan fluorescence at different concentrations of the respective denaturant. We found that FgHET-s(218-289) fibrils were denatured at significantly lower concentrations of both urea and GuHCl than HET-s(218-289) fibrils (Fig. V.2C), indicating that FgHET-s(218-289) fibrils are less stable than HET-s(218-289) fibrils.

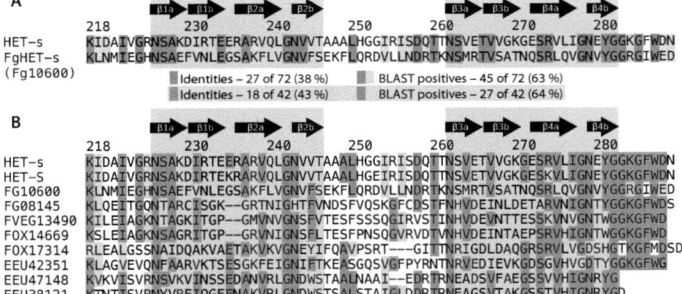

Figure V.1: Sequence alignments of the C-terminal region of HET-s and homologues from different *Fusarium* species. The primary structure of HET-s(218-289) is compared to **A** FgHET-s(218-289) only and to **B** known HET-s homologues. Residues highlighted in green or yellow are identical or have preserved physicochemical properties (BLAST positives), respectively (Moreno-Hagelsieb and Latimer (2008)). The sequence designation of the *Fusarium* homologues corresponds to the GenBank accession numbers. FG10600 and FG08145 are from *F. graminearum*, FOX17314 and FOX14669 from *F. oxysporum*, FVE13490 from *F. verticillioides*, and EEU42351, EEU47148, and EEU38121 from *N. haematococca* (*Fusarium solani*). On top of the alignment, the secondary-structure elements described in the HET-s(218-289) β-solenoid structure in Ref. (Wasmer et al. (2008a)) are given.

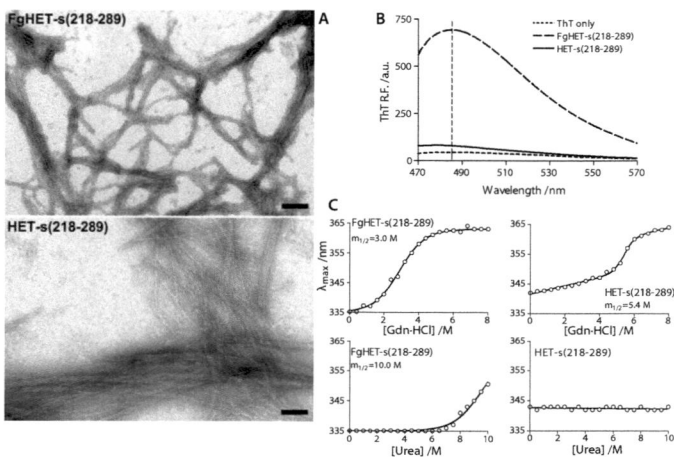

Figure V.2: FgHET-s(218-289) forms amyloid fibrils. **A** Electron micrograph of FgHET-s(218-289) and HET-s(218-289) fibrils (scale bars represent 25 nm). **B** ThT-induced fluorescence of FgHET-s(218-289) fibrils. Note that in contrast to FgHET-s(218-289) fibrils, HET-s(218-289) fibrils do not induce ThT fluorescence. The excitation wavelength was 450 nm and emission was recorded from 470 to 570 nm. ThT and protein concentrations of 25 and 10 µM, respectively, were used. The measurement was performed at both pH 4 and pH 7 and yielded basically identical results (see Fig. X.19 in the appendix). **C** GuHCl (top panels) and urea (bottom panels) induced chemical denaturation of FgHET-s(218-289) and HET-s(218-289) fibrils measured by shift in the maximum emission wavelength of the W287 residue at pH 7. Note that FgHET-s(218-289) are more sensitive than HET-s(218-289) fibrils to both chemical denaturants. As previously reported, no denaturation of HET-s(218-289) fibrils was detected in the presence of urea in these buffer conditions. For the actual UV absorption spectra, see Figs. X.20 and X.21 in the appendix

3.3 *In vitro* FgHET-s(218-289) fibrils seed HET-s(218-289) fibril formation and *vice versa*.

Preformed HET-s(218-289) fibrils are able to suppress the lag phase observed during *in vitro* fibril formation (Balguerie et al. (2003); Sabate et al. (2007)). Therefore we set out to determine whether cross-seeding between FgHET-s(218-289) and HET-s(218-289) is possible *in vitro* and found that preformed FgHET-s(218-289) fibrils are able to accelerate HET-s(218-289) fibril formation and that, *vice versa*, HET-s(218-289) fibrils accelerate FgHET-s(218-289) fibril formation (Fig. V.3). Amyloid fibrils from the unrelated heterologous polypeptide amylin was used as a control and did not show a detectable effect on neither the HET-s(218-289) nor the FgHET-s(218-289) fibril formation rate. Also, seeding with fibrils of full length Ure2p and Sup35 did not accelerate Het-s(218-289) or FgHET-s(218-289) fibril formation(data not shown). We conclude from these observations that *in vitro* cross-seeding between FgHET-s(218-289) and HET-s(218-289) readily occurs.

3.4 Quenched H/D-exchange indicates the location of β-sheets.

In order to determine whether the observed cross-seeding between HET-s(218-289) and FgHET-s(218-289) *in vitro* is related to a similarity in the 3D structures of the two fibrils, we performed quenched H/D-exchange experiments detected by NMR on FgHET-s(218-289) fibrils. H/D-exchange is a sensitive tool for the sequence-specific identification of secondary-structure elements, as backbone amide protons involved in H-bonds are protected from exchange with the solvent and particularly slow exchange is observed for β-sheets. DMSO can be used to solubilize amyloid fibrils into monomers, while preserving the protonation state that was present in the fibrils. This makes the H/D-exchange experiment amenable to a solution NMR analysis by recording fast HMQC spectra of the ^{15}N-labeled protein. This technique has been successfully employed for the structural analysis of amyloid fibrils formed by HET-s(218-289) as well as a number of other amyloidogenic proteins (Ritter et al. (2005); Vilar et al. (2008); Toyama et al. (2007); Hoshino et al. (2002)). Hydrogen exchange in D_2O buffer was followed over 12 weeks. After 4 weeks, the intensities of about 40 % of the resonances were significantly reduced in the spectrum (supplementary Fig. X.13). This demonstrates that the corresponding amide protons had undergone exchange with solvent deuterons, and were therefore no longer detectable in the NMR experiment. The approximately

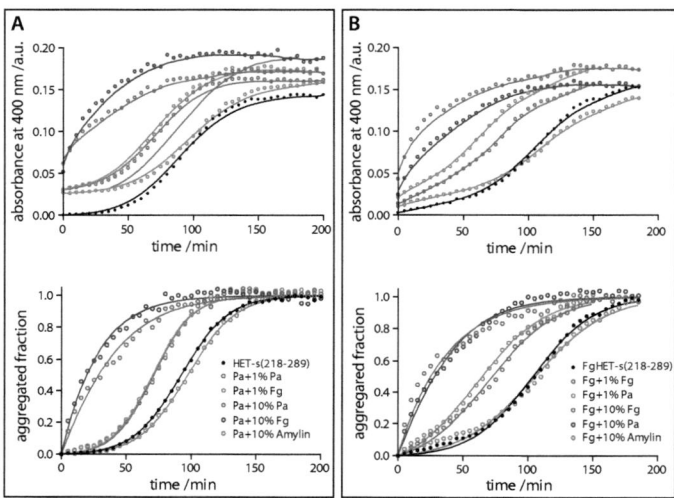

Figure V.3: *In vitro* cross-seeding between FgHET-s(218-289) and HET-s(218-289) fibrils. **A** Aggregation of HET-s(218-289) and **B** aggregation of FgHET-s(218-289) under the influence of different seeds (see legend to the figure). In the top panels, the time courses of the absorbance at 400 nm (raw data) during the fibrillization process are given. The two bottom panels show the normalized data. Note that all FgHET-s(218-289) fibril formation experiments were performed at pH 4.5 to increase the lag phase in spontaneous FgHET-s(218-289) fibril formation.

mono-exponential decay observed for all residues displaying significant hydrogen exchange during the analyzed time interval suggests a well-defined and homogenous structure of the fibrils.

The resulting exchange-rate constants are shown in Fig. V.4C. The backbone amide protons of 5 N-terminal residues, 7 C-terminal residues and residues 246-258 exchanged quickly ($\geq 1.5\,h^{-1}$), i.e. they are only weakly or not protected against the solvent. These residues seem to be not involved in any regular secondary structure elements. We identified four segments displaying very slow exchange rates in the range of $10^{-5}\,h^{-1}$ to $10^{-2}\,h^{-1}$ in good agreement with the exchange-rate constants determined for the β-sheet regions of HET-s(218-289) (Ritter et al. (2005)). The four segments with the highly protected amide hydrogen atoms comprise residues 223-234, 237-245, 259-270 and 273-282, which are thus considered to be involved in hydrogen bonds. In HET-s(218-289), the protected stretches were identified as β-sheets. Within these highly protected regions, residues 243, 265 and 279 show fast exchange. These observations are similar to what has been found in HET-s(218-289), where three of the arcs between the sheets were characterized by a single unprotected residue (Wasmer et al. (2008a); Ritter et al. (2005)).

3.5 Solid-state NMR chemical shifts of FgHET-s(218-289) fibrils reveal high structural similarity with HET-s.

To characterize the rigid parts of the FgHET-s(218-289) fibrils, solid-state NMR experiments employing an initial adiabatic-passage cross-polarization (CP) step (Pines et al. (1973); Hediger et al. (1995)) (from protons to either ^{13}C or ^{15}N) were recorded under magic-angle spinning (MAS). The CP transfer is mediated via the dipolar coupling between the involved nuclei and therefore most effective for rigid parts of the sample, while motion averages out this interaction and therefore quenches the transfer. For example, for HET-s(218-289) this kind of spectra are almost exclusively sensitive for the core region of the amyloid fibrils, i.e. residues 226-249 and 260-282. A CP-MAS solid-state NMR spectrum of U-[^{13}C, ^{15}N] FgHET-s(218-289) amyloid fibrils, a ^{13}C–^{13}C correlation experiment with a 100 ms DARR (Takegoshi et al. (2001); Scholz et al. (2008)) mixing period, is shown in Fig. V.5A (the carbonyl region is shown in Fig. X.14 in the appendix). The spectral resolution is not as good as in HET-s(218-289) (Fig. V.5B), with ^{13}C linewidths in the range of 100 Hz - 200 Hz. For HET-s(218-289) the linewidth ranged between 40 Hz and 100 Hz (comparison at $B_0 = 20.0\,T$, only resolved peaks in the 100 ms DARR spectra of both samples were taken into ac-

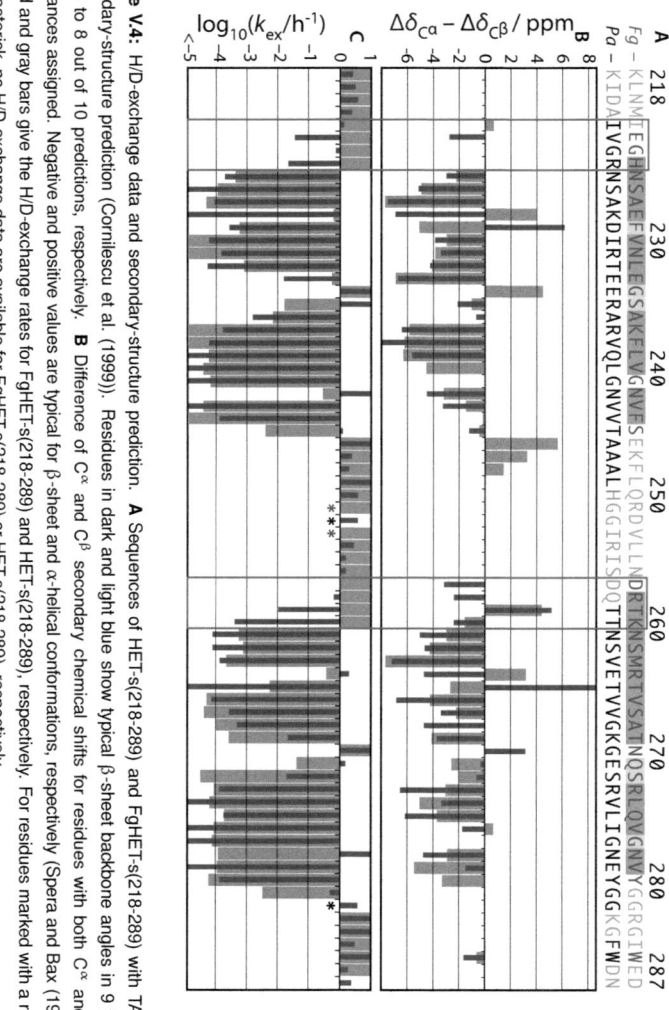

Figure V.4: H/D-exchange data and secondary-structure prediction. **A** Sequences of HET-s(218-289) and FgHET-s(218-289) with TALOS secondary-structure prediction (Cornilescu et al. (1999)). Residues in dark and light blue show typical β-sheet backbone angles in 9 to 10 and 6 to 8 out of 10 predictions, respectively. **B** Difference of C^α and C^β secondary chemical shifts for residues with both C^α and C^β resonances assigned. Negative and positive values are typical for β-sheet and α-helical conformations, respectively (Spera and Bax (1991)). **C** Red and gray bars give the H/D-exchange rates for FgHET-s(218-289) and HET-s(218-289), respectively. For residues marked with a red or black asterisk, no H/D-exchange data are available for FgHET-s(218-289) or HET-s(218-289), respectively.

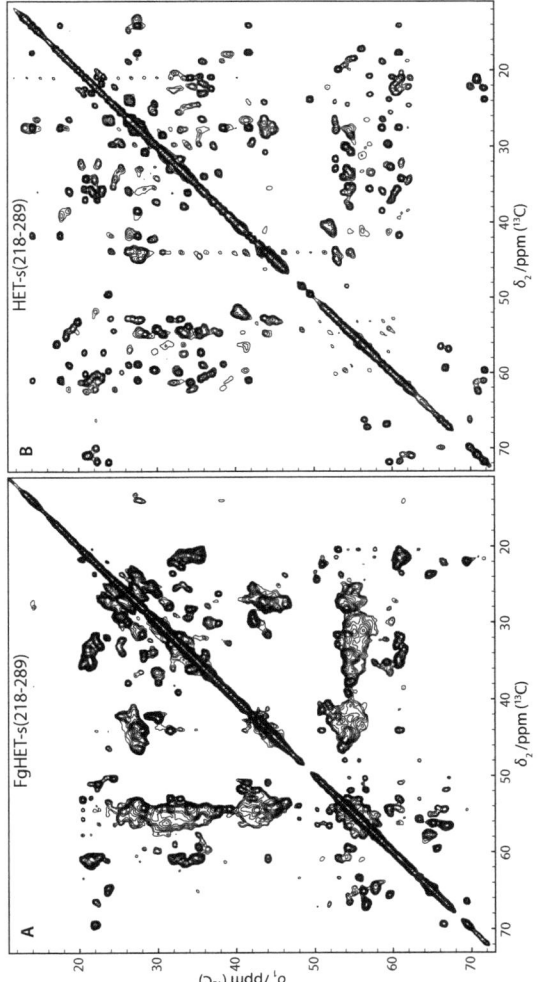

Figure V.5: Aliphatic regions of a PDSD spectrum of U-[^{13}C, ^{15}N]-labeled samples of **A** FgHET-s(218-289) and **B** HET-s(218-289) with 100 ms DARR mixing (Takegoshi et al. (2001); Scholz et al. (2008)). For these short mixing times, short-range (intra-residue and sequential) correlations are dominant. Spectrum (a) was used together with the NCACX and NCOCX spectra (Figs. V.6 and V.7 and Fig. X.15 in the appendix) for sequence-specific assignments. Both spectra were recorded at 850 MHz ^{1}H resonance frequency, 19 kHz MAS frequency, and with 100 kHz SPINAL64 decoupling during t_1 and t_2. Both aliphatic and carbonyl regions of the DARR spectrum of FgHET-s(218-289) are given in Fig. X.15 in the appendix.

count). The increase in linewidth, obtained under otherwise identical conditions, points to a somewhat increased structural heterogeneity of the FgHET-s(218-289) fibrils. Nevertheless, by employing 3D correlation spectroscopy to overcome spectral overlap, we could sequence-specifically assign the resonance frequencies of almost all visible peaks. Heteronuclear correlation spectra, namely NCACX and NCOCX, were recorded with both 2D and 3D acquisition schemes (Detken et al. (2001); Straus et al. (1998); Rienstra et al. (2000); Hong (1999); Siemer et al. (2006b)) and were most useful in the assignment process. An example of the assignment process is shown in Fig. V.6. Additionally a 100 ms DARR spectrum was used to verify backbone assignments and for the assignment of some sidechain atoms. Fig. V.7 shows the 2D N(CO)CX spectrum; both 2D ^{15}N–^{13}C correlation spectra with the assigned peaks labeled are shown in Fig. X.15 in the appendix. Essentially all peaks in the spectrum can be explained by the resonance assignment given in Table X.2 in the Appendix. The details of the resonance assignment process are described in the appendix. Using all recorded spectra jointly, the resonance frequencies of 95 % of the ^{15}N and ^{13}C backbone atoms (N, C', C^α and C^β) within the rigid stretches E223–S246, D258–Y281 and W287 could be assigned sequence specifically (Fig. V.4A, see supplementary Table X.2 for a complete list of assignments).

Due to the strong dependence of the chemical shifts on the polypeptide backbone conformation, these can be used to deduce information about the dihedral angles ϕ and ψ, and to predict the secondary structure. To this aim, the program TALOS (Cornilescu et al. (1999)) was applied to the FgHET-s(218-289) chemical shifts (see Fig. V.4A) and yielded clear predictions for β-sheet conformation (9 or 10 out of 10 database matches) for residues H225-E229, V231-E234, A237-V241, N243-F245 and R259-T270, R274-V277, N279, V280, and strong indications (6 to 8 out of 10 predictions) for β-sheet conformation for F230, G235, S236, Q272, S273 and G278. For residues G224, G242 and N271 the results were ambiguous. Note that TALOS cannot predict the conformation of E223, S246, D258 and Y281, as one neighboring residue of these is not assigned. No residue was predicted to have an α-helical conformation. Additionally, the secondary chemical shifts, meaning the deviation of the chemical shifts from their random-coil value (taken from ref. (Wishart et al. (1995))) were evaluated, which are also indicative for secondary structure elements. In particular, the difference of the C^α and C^β secondary chemical shift, which is positive if a residue is in an α-helical conformation and negative if it's in a β-sheet conformation (Spera and Bax (1991)), has been calculated and

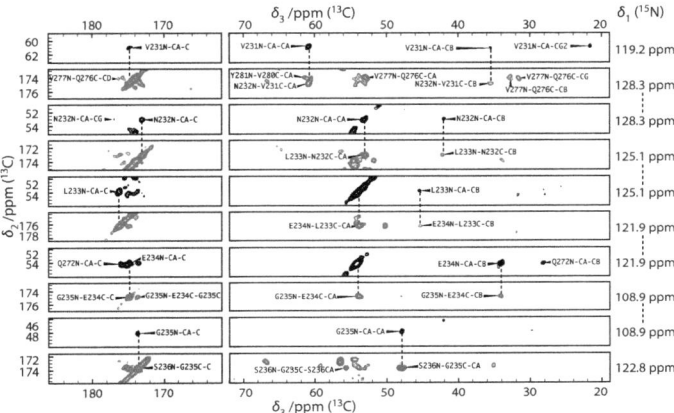

Figure V.6: Strip plots of the 3D NCOCX (gray contours) and 3D NCACX (black contours) spectra (Siemer et al. (2006b)) used for the sequential assignment. The displayed sections illustrate the sequence-specific backbone-resonance assignment for the fragment V231-S236. The spectra were recorded at 850 MHz ^{1}H resonance frequency, 19 kHz MAS frequency, 4 ms N–C CP, 50 ms DARR/MIRROR C–C mixing and 100 kHz SPINAL64 decoupling during t_1, t_2 and t_3.

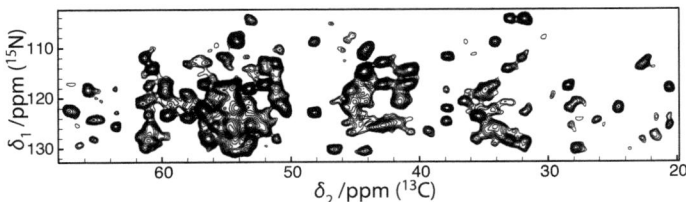

Figure V.7: Aliphatic region of the 2D N(CO)CX solid-state NMR spectrum (Hong (1999)). The spectrum was recorded at 19 kHz MAS, B_0 = 20.0 T, and 50 ms DARR for the C–C mixing period. This spectrum and the N(CA)CX with peak labels are shown in the Appendix.

analyzed. This value, $\Delta\delta_{C^\alpha} - \Delta\delta_{C^\beta}$ is negative for all residues except F230, T260, T266, and N271 (Fig. V.4B, only residues with both C^α and C^β atoms assigned were taken into account). This confirms that FgHET-s(218-289) amyloid fibrils contain almost exclusively β-sheets as secondary structure elements. From the analysis of the chemical shifts and structure of HET-s(218-289) (Wasmer et al. (2008a)) it is known that a single residue with a positive value $\Delta\delta_{C^\alpha} - \Delta\delta_{C^\beta}$ (e.g. K229 and E265) most likely designates the position of a β-arc.

To test for highly dynamical residues, we performed NMR experiments employing an initial H–C INEPT step (Morris and Freeman (1979); Burum and Ernst (1980)) and detection on ^{13}C (Andronesi et al. (2005); Lange and Meier (2008)). In contrast to the CP-type experiments described in the previous section, the INEPT is expected to transfer polarization exclusively for very dynamic moieties that posses sufficiently long transversal relaxation times (T_2). For HET-s(218-289) dynamic residues could be detected that most probably belong to either the N-terminus, a stretch comprising about residues 250-259 or the C-terminus (Siemer et al. (2006a)). The chemical shifts of the observed cross-peaks indicate a random-coil conformation for these parts of HET-s(218-289).

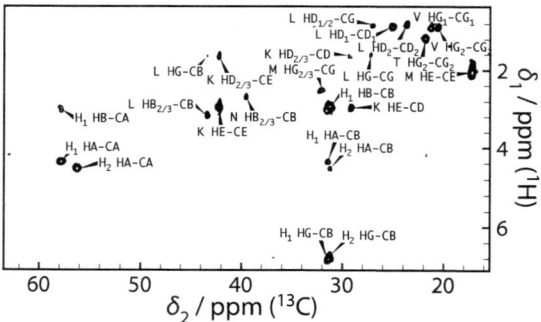

Figure V.8: Aliphatic region of the carbon-detected INEPT experiment with a homonuclear carbon TOBSY transfer performed on U-[^{13}C, ^{15}N]-labeled FgHET-s(218-289) amyloid fibrils. This type of experiment is exclusively sensitive to highly dynamic parts of the protein.

An H(C)C INEPT and an (H)CC INEPT experiment, both with an additional homonuclear ^{13}C–^{13}C TOBSY transfer step (Baldus and Meier (1996); Hardy et al. (2001)) after the initial INEPT, were recorded to facilitate the assignment of the resonances to amino-acid spin systems (Fig. V.8 and Fig. X.16 in the appendix). The INEPT spectra of FgHET-s(218-289) feature only a few detectable resonances that could be assigned to atoms in the side-chains of the amino acids N or D, L, K, M, T and V. Backbone resonances were only found for two H spin-systems, most likely arising from the C-terminal H_6-tag. In comparison to HET-s(218-289) (Siemer et al. (2006a); Lange and Meier (2008)), significantly fewer signals are observed for FgHET-s(218-289) which indicates that less residues are flexible enough to show up in this type of experiments. The chemical shifts of the assigned resonances closely resemble the random-coil values (Wishart et al. (1995)).

3.6 Cross-seeded fibrils adopt a similar structure as unseeded.

The electron micrographs of the seeded fibrils have very similar features as those of the unseeded fibrils for both HET-s(218-289) and FgHET-s(218-289) (Fig. X.17 in the Appendix) and the FgHET-s(218-289) showed florescence with ThT, also if seeded with HET-s(218-289). The 100 ms DARR solid-state NMR spectrum of FgHET-s(218-289) fibrils seeded by preformed HET-s(218-289) (Fig. X.18 in the Appendix) shows that the seeded fibrils exhibits the same chemical shifts as the unseeded ones and therefore also have the same structure. Nevertheless, some differences are found, in particular a broader lineshape for the seeded sample, indicative of an increased disorder or polymorphic behavior (Fig. X.18). Detailed investigations of this phenomenon are presently under way.

4 Discussion and Conclusions

4.1 Structural comparison to HET-s(218-289).

The secondary chemical shifts as a function of the primary structure, as extracted from the solid-state NMR assignment of FgHET-s(218-289), closely resemble that of HET-s(218-289) (Fig. V.4b). This implies that FgHET-s(218-289) contains β-sheet elements in almost the same positions as HET-s(218-289). The few residues with positive secondary chemical-shift differences $\Delta\delta_{C^\alpha} - \Delta\delta_{C^\beta}$ (F230, T260, T266 and N271) most likely indicate the positions of β-arcs connecting sequentially adjacent β-strands (as also seen in HET-s(218-289)) (Wasmer et al. (2008a); Ritter et al. (2005))). The lower protection from H/D-exchange of these residues confirms this and indicates β-arcs at G235-S236, N243, R265, N271-Q272 and

N279. The H/D-exchange data are more complete as no chemical shift analysis was performed for glycine residues that happen to be particularly abundant within (or just before) a β-arc. The fact that each of the β-arcs has a partner at ±36 residues ([F230, T266], [G235, N271], [N243, N279]) suggests that the two pseudo-repeats 223-245 and 259-281 form parallel β-sheets with one another as seen in HET-s(218-289).

The most obvious difference to HET-s(218-289) is the appearance of additional rigid residues in FgHET-s(218-289), namely 223, 224 and 258-260, that could form a very short β-sheet and maybe a connecting β-arc (green boxes in Fig. V.4). The reason for this might be found in the two oppositely charged residues E223 and R259, separated by exactly 36 residues in the FgHET-s(218-289) sequence and therefore partners in a hypothetical additional N-terminal β-sheet. The sidechains of these two residues may form a salt-bridge and thereby stabilize the β-sheet. For HET-s(218-289) no such interaction is conceivable as valine and glutamine are the respective residues at positions 223 and 259 and accordingly residues 222-225 and 258-262 are not in a β-sheet. This finding is supported by the fact that H/D exchange is very fast here (Ritter et al. (2005)).

In HET-s(218-289) residues 247 to 261 are only weakly protected from H/D-exchange and only the beginning and end of this stretch are visible in CP-type solid-state NMR experiments, indicating a high degree of dynamics for residues in the center of the loop (Siemer et al. (2006a)). Indeed, in HET-s(218-289) these residues are observable in INEPT experiments. For FgHET-s(218-289), on the other hand, no residues flexible enough to show backbone atoms in INEPT spectra were detected in the loop pointing towards a shorter, less flexible loop. All residues detected in the INEPT experiment show nearly the average chemical shift values (according to the BMRB (Ulrich et al. (2008)); except for one His, which may be located in the C-terminal His$_6$ tag) which indicates that these residues are indeed flexible and not part of a highly dynamic but folded domain.

A remarkable difference between FgHET-s(218-289) and HET-s(218-289) occurs in the core region that otherwise seems to have a highly conserved structure between the proteins. The first β-arc is positioned at residues K229-D230 and E265-T266 for HET-s(218-289) (Wasmer et al. (2008a)). In FgHET-s(218-289) the position where the secondary chemical shifts deviate significantly from the values expected for a β-sheet is shifted by one residue, while the H/D-exchange data show fast exchange at the same positions (see discussion below). This behavior could be explained by the fact that multiple types of two-residue β-arcs exist. Whereas a so called ab arc (Hennetin et al. (2006)) occurs at this position in HET-s(218-289), a bl arc, the most abundant form, could be present in FgHET-s(218-289). This arrangement would show basically identical side-chain arrangement (inside vs. outside) but different backbone angles for E229 and F230 (R265 and T266), i.e. a change in the consecutive dihedral angles ψ_{229} and ϕ_{230} (ψ_{265} and ϕ_{266}) of about 180° each. This arrangement could explain the observed differences in the secondary chemical shifts (Wishart and Nip (1998)).

The finding that E229, positioned at this β-arc, is highly protected from H/D exchange, while the expected partner R265 has a high H/D-exchange rate can only be explained by differences in the H-bonding pattern occurring in the β-arcs at these positions. A similarly high protection of a residue within a β-arc has been observed for HET-s(218-289), where N243, connecting β2a and β2b, displays relatively fast hydrogen exchange, while the corresponding residue N279 connecting β4a and β4b is fully protected. A more detailed explanation has to await the full structure determination.

Another notable difference between the two fibrils is observed at the end of the first pseudo-repeat, around residues 246-249. In HET-s(218-289) three alanine residues occupy positions 247-249 that, despite being unprotected from H/D-exchange, are visible in CP-type spectra and exhibit chemical shifts typical of α-helices (Ritter et al. (2005)). In FgHET-s(218-289), already the primary structure of this part, as well as of the whole flexible loop, does not bear any resemblance to HET-s(218-289). Also, there is no evidence in the CP spectra of the corresponding residues E247, K248 and F249 and therefore these are likely to be dynamically disordered.

In addition to the two pseudo-repeat regions, there is only a single additional amino-acid residue assigned in the CP-type spectra of FgHET-s(218-289). W287 is a conserved residue in most HET-s homologues (Fig. V.1b) and is also present in HET-s(218-289) itself. The Tryptophan side-chain has been found to make contact with residues in β2a and β4a, one of the β-sheets confining the hydrophobic core region in HET-s(218-289) (Van Melckebeke et al. (2010)) and this residue is necessary for the prion infectivity and in vivo aggregation of HET-s (Sandra Cescau, Sven J. Saupe, unpublished results). This residue is, except for a few resonances of the side-chain of the neighboring F286 in HET-s, the only observable moiety of the C-terminus in CP-type NMR spectra of both FgHET-s(218-289) and HET-s(218-289). Therefore, it has to be at least partially immobilized in both protein fibrils underlining its importance for fibril formation.

Assuming the same fold for the core region comprising the two stretches 226-244 and 262-280 of FgHET-s(218-289) and HET-s(218-289) as depicted in Fig. V.9, the resulting organizations of hydrophobic and hydrophilic side-chains is quite similar in FgHET-s(218-289) and HET-s(218-289). The hydrophobic core of the FgHET-s(218-289) fibrils has to accommodate only 1 polar but 11 non-polar (2 polar and 10 non-polar in HET-s) amino acid side-chains, whereas those pointing outside are, except for one (F230), either polar or charged making the model rather appealing.

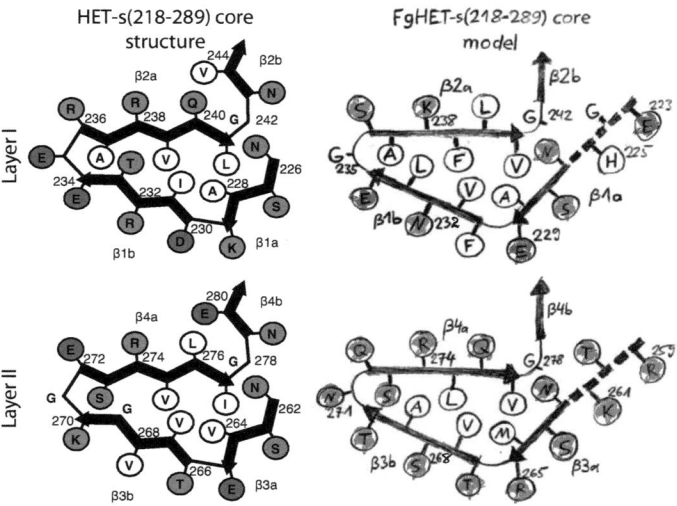

Figure V.9: Schematic representation of the two layers of the structure of the well-ordered hydrophobic core of PaHET-s(218-289) amyloid fibrils (Wasmer et al. (2008a)). This fold would explain all currently available data and is compatible with the primary structure of FgHET-s(218-289). Circles representing positively or negatively charged residues are colored blue or red, respectively; polar but uncharged residues are green and nonpolar residues are white.

Another stabilizing feature observed for HET-s(218-289) fold is the possible formation of three inter-β-strand salt-bridges between the oppositely charged side chains of K229-E265, E234-K270 and R236-E272 that may form both intra- and inter-molecularly (Lange et al. (2009)). Only one of these is reproduced in the FgHET-s(218-289) homology model (Fig. V.9), namely between E229 and R265, with the charges inverted compared to HET-s(218-289). An additional pair of oppositely charged residues, E223-R259 may form a second salt-bridge and might be the stabilizing element leading to the prolongation of β-sheet 1a/3a or the addition of a short β-sheet in FgHET-s(218-289). Whether the reduction of stability of the FgHET-s(218-289) fibrils (Fig. V.2C) is indeed a direct consequence of the lower amount of possible salt-bridges remains to be determined. The broader spectral lines in the solid-state NMR spectra of FgHET-s(218-289) could be attributed to a higher degree of conformational disorder which might be related to a lower fibril stability or the smaller number of specific interactions that must be met by the structure to avoid significant energetic penalties.

The comparison of HET-s(218-289) and the structural model of FgHET-s(218-289) also allows to propose an explanation for the different behavior of the two proteins regarding the induced ThT-fluorescence. It has been proposed that ThT binds amyloids in the "channels" running along the fibril axis delimited by the side chains of the i and $i + 2$ residues of each β-strand (Krebs et al. (2005)). ThT is positively charged and thus positively charged residues hinder ThT binding. All "channels" formed in HET-s(218-289) are lined by at least one positively charged residue while, in contrast, in the FgHET-s(218-289) model several of the accessible "channels" are free of basic residues. This observation might explain why in spite of their overall structural similarity, FgHET-s(218-289) fibrils robustly induce ThT-fluorescence while HET-s(218-289) fibrils fail to do so.

4.2 Evolutionary conservation of the β-solenoid fold.

We have shown that FgHET-s(218-289) has the ability to form amyloids which are structurally highly similar to the HET-s(218-289) β-solenoid fold. The apparent structural similarity of the FgHET-s(218-289) and the HET-s(218-289) fibrils may seem surprising given the relatively low sequence identity of the two constructs (38 %). A closer look however reveals that most of the conserved regions lie within the rigid and well-defined parts, the two pseudo-repeat regions 222-247 and 258-283 which have 43 % sequence identity (green in Fig. V.1A). The non-conserved residues with similar physico-chemical properties (BLAST "positives", yellow in Fig. V.1) are however scattered over the whole sequence. The sequence alignment shown in Fig. V.1B reveals that the conservation of residues that play a key role in the β-solenoid fold of HET-s(218-289) extends to other identified het-s homologues in *Fusarium* species. For instance, the asparagine residues which form two ladders along the fibrils axis in HET-s(218-289) (N226, N243 and N262, N279), are conserved in all homologues. The same is true for the G240 and G278 residues allowing the formation of the β-arc leading into the fourth strand pointing away from the triangular hydrophobic core. All inward-facing hydrophobic residues in HET-s(218-289) (A228/V264, I231/V267, V239/V275, L241/I277) also show conservation in all *Fusarium* homologues. Finally, the glycine-rich C-terminal loop containing W287, which has been found to make contact with residues in β2a and β4a (Van Melckebeke et al. (2010)) is also conserved in many homologues (with the exception of two *Nectria* sequences). These observations strongly suggest that a selective pressure to maintain the ability to form this β-solenoid structure, including the C-terminal residues, is operating. The estimated divergence time between P. anserina and *F. graminearum* is roughly in the range of 400 million years. During this period the sequences of HET-s and FgHET-s have highly diverged, but in a way that allows to conserve amino acid positions important for the formation of the β-solenoid fold.

4.3 Structural similarity accounts for the cross-seeding in between HET-s(218-289) and FgHET-s(218-289).

Amyloid cross-seeding between HET-s(218-289) and FgHET-s(218-289) occurs in spite of a considerable divergence of the primary sequence. Our structural analysis provides a simple explanation for this cross-seeding ability: the actual structural similarity between HET-s(218-289) and FgHET-s(218-289) fibrils. This result suggests that amyloid templating is possible at moderate levels of sequential identity if structural similarity is ensured. Some indications of an increase in structural disorder are found for the seeded fibrils and this observation will be followed up.

5 Summary

We conclude that, on a structural level, FgHET-s(218-289) is closely related to HET-s(218-289). In particular, the hydrogen exchange rates and the NMR chemical shifts indicate that the triangular hydrophobic core is conserved and that the major elements that additionally stabilize the core of the fibrils in HET-s(218-289), namely at least 21 hydrogen bonds per molecule and one of the three salt bridges in HET-s(218-289), are conserved. The similarity of the structural models could explain the observation that *in vitro* cross-seeding is possible, even though the

HET-s(218-289) and FgHET-s(218-289) proteins only exhibit moderate levels of sequence identity. On a more general level, our study illustrates the fact that two amyloid proteins sharing 38 % sequence identity can adopt highly similar structures. It is largely documented that homology levels in the range of 30 % can lead to similar structures in soluble proteins (Chothia and Lesk (1986); Ginalski (2006)). Here we present an example in which the same principle is applicable to amyloid structures despite the known tendency of amyloids to form different polymorphic forms and the fact that even point mutations have been shown to lead to completely different structures, e.g. parallel and anti-parallel β-sheets (Tycko et al. (2009)).

V A HET-s Homologue—FgHET-s

Chapter VI

The Structure of FgHET-s

This work was done in collaboration with Peter Güntert and is being prepared for publication. Peter implemented the use of solid-state NMR data in the newest version of the CYANA software.

1 Introduction

Prions, infectious entities composed solely of protein (Prusiner et al. (1998)), are the cause of a number of diseases in mammals. However, prions have also been identified in fungi (Wickner et al. (2007)). As model systems, these allow us to study processes of interest, like prion propagation, more closely. HET-s is a prion of the filamentous fungus *Podospora anserina* and is involved in a self/non-self recognition reaction termed heterokaryon incompatibility (Coustou et al. (1997)). HET-s forms fibrils in vitro that are infectious and have been structurally characterized (Wasmer et al. (2009b)). They feature a well-defined triangular hydrophobic core formed by the C-terminal residues 218-289 while the rest of the protein is dynamically disordered. The prion domain of HET-s in isolation, HET-s(218-289), exhibits the same fold as in the full-length protein (Wasmer et al. (2009b)). It is highly ordered and free from any form of polymorphism, which allowed for an atomic-level structure determination by solid-state NMR (Wasmer et al. (2008a); Van Melckebeke et al. (2010)). The fold of HET-s(218-289) can be described as a β-solenoid, in which one molecule contributes two windings.

A distant homologue of HET-s has been found in the filamentous euascomycete *Fusarium graminearum*, a prominent wheat, barley, maize and oat pathogen (Parry et al. (1995)) and it has been termed FgHET-s. FgHET-s has the same length as HET-s (289 residues) and displays an overall sequence identity of 50 %, but only 38 % for the prion domain (residues 218-289). FgHET-s(218-289) also forms amyloid fibrils in vitro and these are able to efficiently cross-seed fibril formation of HET-s(218-289) (Wasmer et al. (2010)). FgHET-s(218-289) has been characterized biochemically, and hydrogen/deuterium (H/D) exchange data and an almost complete solid-state NMR assignment has been obtained. From this information, a structural model of FgHET-s(218-289) based on homology to HET-s(218-289) has been proposed (Wasmer et al. (2010)).

In the following, we describe the atomic-resolution structure determination of the FgHET-s(218-289) fibrils using restraints derived from solid-state NMR spectra and H/D-exchange data. As FgHET-s(218-289) exhibits broader solid-state NMR lines and also a somewhat less advantageous distribution of chemical shifts than HET-s(218-289), no unambiguous distance restraints could be obtained. Based on the latest version of CYANA (Guntert et al. (1997)) and the structure calculation protocol for amyloid fibrils established for HET-s(218-289) (Van Melckebeke et al. (2010)) we established a structure calculation implementation that automatically rates and assigns ambiguous distance restraints (network anchoring) from solid-state NMR spectra. The new protocol works completely automatically. The input data was recorded in only 20 days on 3 protein samples with basic (cheap) labeling schemes. The structure of FgHET-s(218-289) could be obtained at high resolution for the hydrophobic core region. This allowed us to compare the structures of HET-s(218-289) and FgHET-s(218-289) in atomic detail and confirm the high evolutionary conservation of the HET-s core fold. We also find a number of differences that are however either marginal or outside the hydrophobic core region.

2 Materials and Methods

2.1 Sample Preparation

The region corresponding to the prion domain of HET-s, i.e. residues 218 to 289, of FgHET-s with the addition of one N-terminal methionine and a C-terminal histidine$_6$ tag was recombinantly expressed in *E. coli*. The following purification and fibrillation was performed as described for HET-s(218-289) (Balguerie et al. (2003); Wasmer et al. (2010)). In order to obtain a high signal intensity in the NMR spectra and therefore a large number of distance restraints it is optimal to record spectra on homogeneously labeled samples with all carbon and nitrogen atoms carrying a spin 1/2 (i.e. ^{13}C and ^{15}N). However, in spectra of such homogeneously labeled samples, it is impossible to distinguish between intra- and intermolecular contacts and we used two additional labeling schemes to accomplish this task: 'Dilute' samples, in which only 25 % of the molecules are labeled and 'mixed' samples that are fibrillated from a 1:1 mixture of U-^{13}C and U-^{15}N FgHET-s(218-289) (Wasmer et al. (2008a); Van Melckebeke et al. (2010)). All NMR spectra used in this study for the structure calculation were performed on one of these three samples (homogeneous, dilute or mixed).

In contrast to the work on HET-s(218-289), we did not use 2-^{13}C or 1,3-^{13}C glycerol grown samples (LeMaster and Kushlan (1996)) and no experiments with a

proton-driven spin diffusion (PDSD, DARR) mixing step were performed.
All solid-state NMR samples were centrifuged into 3.2 mm Bruker magic-angle spinning (MAS) rotors at 200,000 g, which were sealed using epoxy in order to prevent sample dehydration during the measurements.

labeling	mixing scheme	mixing time	rec. time	assign tolerance (dim 1 / dim 2)	peaks picked	initial upper limit (upl)
homo-	PAR	8 ms	60 h	0.5 / 0.4 ppm	1263	7.0 Å
geneous	PAR	16 ms	74 h	0.5 / 0.45 ppm	795	9.0 Å
dilute	PAR	10 ms	92 h	0.5 / 0.4 ppm	387	7.0 Å
mixed	PAIN	4 ms	112 h	0.8 / 0.4 ppm	56	6.0 Å
	NHHC	250 µs	66 h	0.8 / 0.4 ppm	25 – 2	4.5 Å
total			17 d		2526 – 2	

Table VI.1: List of spectra used in the structure calculation of FgHET-s(218-289). The column tagged "labeling" indicates the sample used for the experiments listed. "Homogeneous" samples are fully U-[^{13}C, ^{15}N] labelled, "dilute" samples contain 25 % U-[^{13}C, ^{15}N] labelled and 75 % unlabeled molecules, and "mixed" samples consist of 50 % U-^{13}C and 50 % U-^{15}N labelled molecules. We expect the differently labelled proteins to arrange stochastically within the fibrils.

2.2 Solid-state NMR

All NMR spectra evaluated in the course of the structure calculation were performed on a Bruker Avance II+ wide-bore NMR spectrometer operating at a static magnetic field of 20 T (850 MHz ^{1}H resonance frequency) using a Bruker 3.2 mm triple-resonance probe. The MAS frequency was actively stabilized at 19.00 kHz and the sample temperature was 3 °C for all experiments. SPINAL-64 proton decoupling (Fung et al. (2000)) with an RF amplitude of 100 kHz was applied during t1 and t2 evolution periods. All cross-polarization (CP) and third-spin-assisted recoupling (TSAR) (Lewandowski et al. (2007)) steps were optimized to obtain optimal polarization transfer between the desired nuclei. Table VI.1 lists all recorded spectra, i.e. 2 proton-assisted recoupling (PAR) spectra on homogeneously U-[^{13}C, ^{15}N] labeled FgHET-s(218-289) fibrils with 8 ms and 16 ms mixing time, respectively, a PAR spectrum on dilute U-[^{13}C, ^{15}N] labeled fibrils with 10 ms mixing and a 4 ms proton-assisted insensitive nuclei (PAIN) CP and an NHHC spectrum with 250 µs mixing on the mixed ^{13}C/^{15}N labeled sample. The peak-picking was performed automatically in Sparky 3.115 (T. D. Goddard and D. G. Kneller, University of California, San Francisco) for all spectra. Regions containing the spinning sidebands of strong diagonal signals in the homonuclear carbon spectra were spared out.

2.3 Implementation of the Structure Calculation

The structure calculation was performed on a trimer of loosely connected (121 residues linker) FgHET-s(218-289) molecules, which is the minimal entity to account for all conceivable features of a 1-dimensional repetitive structure of identical molecules with identical adjacencies (Van Melckebeke et al. (2010)). The fact that all molecules within the fibril are folded identically can be directly derived from the observation of a single set of chemical shifts. This is implemented by symmetry restraints that ensure that each interatomic intramolecular distance is identical in all three molecules. (In detail, the difference between 'one' distance in two different molecules is contributing an energy term in the simulated annealing process).

For the structure calculation we used a version of CYANA 3.0 (Guntert et al. (1997)) that was adapted for solid-state NMR such that it allowed to use restraints between heavy atoms (previously only protons). The input for a calculation consisted of the chemical shift assignments, a peak-list from each spectrum, H-bonds, dihedral angle restraints from TALOS, Ramachandran and side-chain rotamer dihedral angle restraints as given in detail below. Each peak-list also contains information on how to interpret it, i.e. the types of atoms involved and if the restraints to be derived should be intra-, intermolecular or ambiguous. In the following the details of the structure calculation process are described.

The initial values for the tolerances in the peak assignment process were set to values slightly above the actual spectral resolution (see Table VI.1). The distances to be used as restraints in the first cycle of the calculation were set to the values given in Table VI.1.

A structure calculation comprises 7 cycles, throughout which CYANA tries to obtain the correct assignments of the peaks even if these were initially highly ambiguous. In cycle 1 assignment possibilities of peaks are taken into account only if the resulting restraints are additionally supported by others (Network anchoring). How strong this support has to be can be adjusted and we did so to account for the characteristics of the solid-state NMR data, i.e. that most peaks have numerous assignment possibilities. The parameters to work best with the data used herein proved to be $p = 0.09 \times c$ and $q = 0.4$ (cycle 1) but the final outcome of the structure calculation is robust, i.e. not very sensitive to these values.

In each of the 7 cycles, 100 structures were annealed and 10 of these were analyzed and used to assign peaks for the following cycle. The annealing of each structure was done in 200,000 steps, leading to a CPU time of 1-3 h per structure. All 100 calculations were carried out in parallel on the Brutus cluster at ETH Zurich (www.brutus.ethz.ch). As each cycle needs to wait for the slowest calculation to finish and then collects and evaluates the resulting structures, a complete run of 7 cycles could be performed in 1 day. (However, multiple calculations with different input data can be run in parallel.)

3 Results

3.1 Spectra and Distance Restraints

The distance restraints for the structure calculation were extracted from the 5 spectra listed in Table VI.1. The 2 homonuclear 2-dimensional carbon spectra with the PAR mixing scheme (De Paepe et al. (2008)) on the homogeneously labeled sample yielded 1263 and 795 peaks for a mixing time of 8 ms and 16 ms, respectively (Fig. VI.1 A). It should be noted, that the 16 ms spectrum shows basically no cross-peaks between both aromatic and carbonyl resonances and the aliphatic carbons, which is most likely due to the specific experimental conditions used (RF field amplitudes during the PAR transfer). On the dilute sample, one homonuclear carbon spectrum with 10 ms PAR mixing was recorded. As only part of the molecules are labeled here, the signal intensity and therefore also the number of peaks is reduced. However, as the intensity for cross-peaks arising from intermolecular contacts is reduced by a factor 4 with respect to the intramolecular ones, we can assign all signals to the latter with no or very few errors. Finally, a PAIN spectrum (Fig. VI.1 B) (Lewandowski et al. (2007)) and an NHHC spectrum (Lange et al. (2002)) were recorded to probe the intermolecular interface of the FgHET-s(218-289) fibrils. Both spectra show a relatively small numbers of peaks (56 and 25, respectively), but here each peak originates from an intermolecular long-range contact, i.e. all peaks in these two spectra contain meaningful structural information.

On the other hand, for the experiments on homogeneous or dilute samples part of the peaks are structurally meaningless diagonal peaks or short-range contacts and

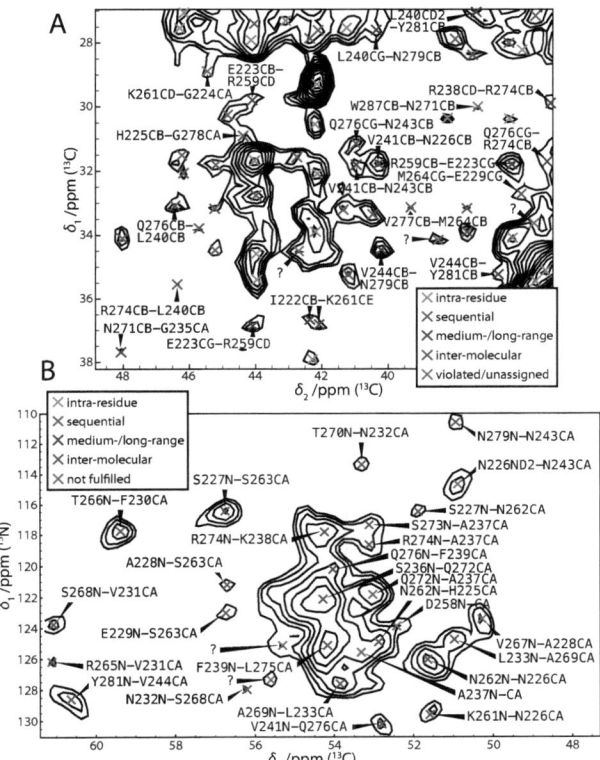

Figure VI.1: **A** Section of the 8 ms PAR spectrum on U-[^{13}C, ^{15}N] labeled FgHET-s(218-289) and **B** section of the 4 ms PAIN spectrum on mixed ^{13}C/^{15}N labeled FgHET-s(218-289). The peaks are colored according to their assignment in cycle 7 of the CYANA structure calculation and the assignment is given. Short-range peaks (green) are not labeled. **A** Three peaks in this region are not explained by the final structure ("?") while two are not assigned in the final cycle but can be explained by the obtained structure (violated by 1-2 Å). **B** In the intermolecular PAIN spectrum, most peaks leading to restraints that are not fulfilled in the final calculation can be explained by contacts that are 1-2 Å longer than 6 Å. These have labels but are colored red. Two weak peaks can not be explained and are labeled by "?".

the number of observed signals is not a good measure for the information content of each spectrum. The real 'value' of a spectrum can only be seen from its impact on the structure calculation itself (Van Melckebeke et al. (2010)).

3.2 Complementary Restraints—H/D exchange and TALOS

As for HET-s(218-289), complementary structural information in the form of hydrogen/deuterium exchange rates measured by solution NMR was available for FgHET-s(218-289) (Wasmer et al. (2010)). We implemented this data in the form of H-bonds between the 2 pseudo-repeats of residues 223-245 and 259-281, for all residues with $k_{ex} < 0.1\,h^{-1}$ in an ambiguous way, such that each H-bond can be either intra- or intermolecular. However, the structure calculation has a practically identical outcome with a minimal set containing only 1 H-bond per β-strand.

In addition to the distance restraints derived from the peak-lists and the H/D-exchange data, TALOS dihedral angle restraints (Cornilescu et al. (1999)) were derived from the chemical shifts of the assigned atoms. The angles proposed by TALOS were kept only in cases where 9 or 10 out of 10 database matches agreed for a specific triplet of residues (see figure 4(a) in Wasmer et al. (2010)) and were completely discarded otherwise. Furthermore, angle restraints to favor allowed Ramachandran plot regions ("ramaaco") and angle restraints for staggered side-chain rotamers ("rotameraco") were created by CYANA 3.0 (Guntert et al. (1997)).

3.3 The Structure Calculation

In the case of FgHET-s(218-289) the distribution of chemical shifts and the spectral resolution do not allow for a single peak to be assigned unambiguously to a long-range correlation. This was quite different for HET-s, where 39 such peaks could be found and the initial structure calculation is based on them (Wasmer et al. (2008a)). The calculation of cycle 1 is therefore relying exclusively on ambiguous distance restraints that are rated and edited using network anchoring. The optimal value for the probability threshold for the network anchoring is considerably lower than for solution NMR data, but the calculation is robust with respect to it, i.e. the cycle 1 structures vary slightly when changing the parameter but the final cycle yields the same fold over a vast range of values. Or, in other words, if the initial cycle yields a reasonable fold, CYANA succeeds in finding the correct assignment possibilities and the calculation always converges to the same fold (with the exception of one structural detail, see below).

The RMSD for FgHET-s(218-289) was always calculated for an entity consisting of residues 223 to 245 and 259 to 281. With the structure calculation protocol described in the methods section, already the initial cycle yields a quite well-defined fold with a backbone RMSD of 0.9 Å, but a relatively high energy of 165 Å^2 indicates that some restraints are violated. In the 6 following cycles (see Table VI.2), the number of assignment possibilities per restraint are reduced from 24.87 to 6.10 (on average per peak) (see Fig. VI.2), which results, as expected, in an improvement of the RMSD, as apparent from the bundle of the 10 lowest-energy structures in Fig. VI.3. The number of assigned peaks, however, does not change significantly (from 2082 to 2039) throughout the cycles.

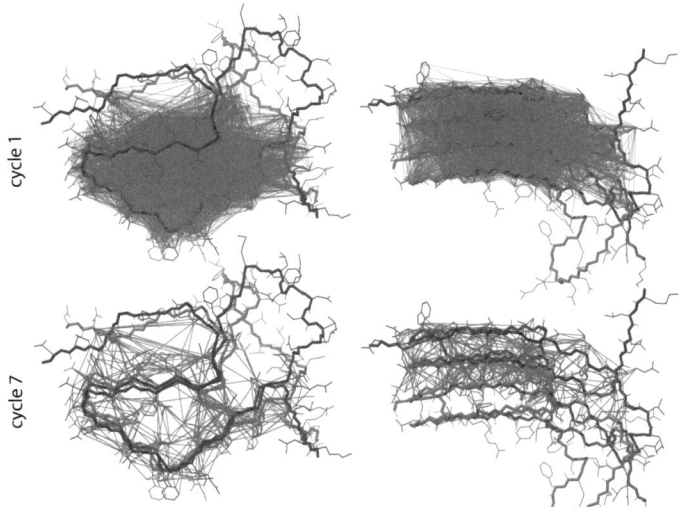

Figure VI.2: Distance restraints in the initial (top) and the final cycle (bottom). Top and side views of two FgHET-s(218-289) molecules within the fibrils are given. The intramolecular restraint assignments are shown in the upper (dark blue) molecule only, while intermolecular restraints are plotted between the two molecules. In the final cycle all restraints are unambiguous (1 assignment per restraint).

For the final cycle, CYANA choses the one most probable assignment possibility

	cycle 1	cycle 7
selected	2501	2501
in 8 ms PAR (hom)	1263	1263
in 16 ms PAR (hom)	795	795
in 4 ms PAIN (mix)	56	56
in 10 ms PAR (dil)	387	387
assigned	2082	2039
unassigned	419	462
without assignment possibility	215	355
with violation below 0.5 Å	204	40
with violation between 0.5 and 3.0 Å	0	60
with violation above 3.0 Å	0	7
in 8 ms PAR (hom)	253	285
in 16 ms PAR (hom)	92	112
in 4 ms PAIN (mix)	23	13
in 10 ms PAR (dil)	51	52
with diagonal assignment	165	165
Cross peaks		
with off-diagonal assignment	1917	1874
with unique assignment	0	125
Peaks with increased upper limit		161
Peaks with decreased upper limit		0

Table VI.2: Structure calculation statistics of cycles 1 and 7.

Results 3

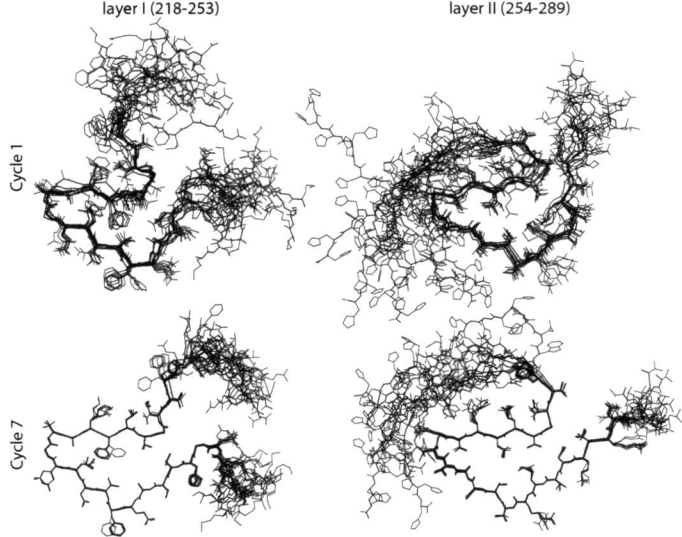

Figure VI.3: 10 lowest energy structures out of 100 calculated structures. On top, the structures obtained in cycle 1, at the bottom, the structures obtained in the final cycle (cycle 7) are shown. For sake of clarity, the first half of FgHET-s(218-289), residues 218 to 253, is shown in the left panels, the second half, residues 254 to 289 plus the C-terminal H_6-tag, is shown in the right panels. Note that the structures of β_{1a}/β_{3a} and of Arc 3 do change slightly between cycle1 and cycle 7.

for all assignable peaks (see below for details on the non-assignable ones), while in cycle 6, each peak still had on average 6.1 assignments. The outcome of the structure calculation however is almost identical in cycles 6 and 7, and a bundle of structures with an RMSD of 0.12 Å (0.71 Å for heavy atoms) is obtained (Fig. VI.3). Also, in the context of amyloid fibrils, there can and will always be peaks that—even with the knowledge of a high-resolution structure—can never be attributed to a single interatomic distance. For example consider an intra-β-sheet contact between backbone-atoms of the same type (e.g. C^α-C^α) derived from a homogeneously labeled sample (i.e. it can be either intra- or intermolecular). Such a restraint can always be fulfilled intra- and intermolecularly simultaneously, as the respective distances are almost identical. Therefore, we expect the assignments made in cycle 6 to be a more accurate representation of the actual provenance of the observed cross-peaks.

In the final cycle 469 peaks (plus 2 from the NHHC) are not assigned and therefore also not explained by the structure. However, 215 peaks have no possible resonance assignment at all (see Table VI.2 cycle 1), which means that they are, in at least one dimension, located at frequency positions not associated with any assigned resonance. These are therefore either spectral artifacts or, and we consider this to be more likely, belong to unassigned spin systems. A good example of one or more unassigned spin-systems is observable in the 16 ms PAR spectrum on the homogeneous sample, in the Isoleucine side-chain region (at 10-12 ppm and around 16 ppm) (Fig. VI.4). About 20 unassigned peaks are visible in this part of the spectrum that do however not lead to wrong restraints and therefore do not influence the structure calculation adversely.

The rest of the unassigned peaks (254) have possible assignments that do not fit the final structure. One hundred of those have assignment possibilities that could be fulfilled by increasing the corresponding distance restraint by less than 3 Å. These peaks could be explained by more favorable transfer conditions for the TSAR transfer (Scholz et al. (2007)) due to the spin topology in certain parts of the protein. However, these and all remaining peaks can also be explained by missing resonance assignments. In a very similar system, HET-s(218-289), it has been proven difficult to obtain a complete resonance assignments as regions of the protein in amyloid fibrils exhibit dynamics that impair the CP transfer steps used in most experiments. In HET-s(218-289) the initial assignment was complemented during many years based on a large amount of experiments (Ritter et al. (2005); Siemer et al. (2006b); Wasmer et al. (2008a); Van Melckebeke et al. (2010)). With the

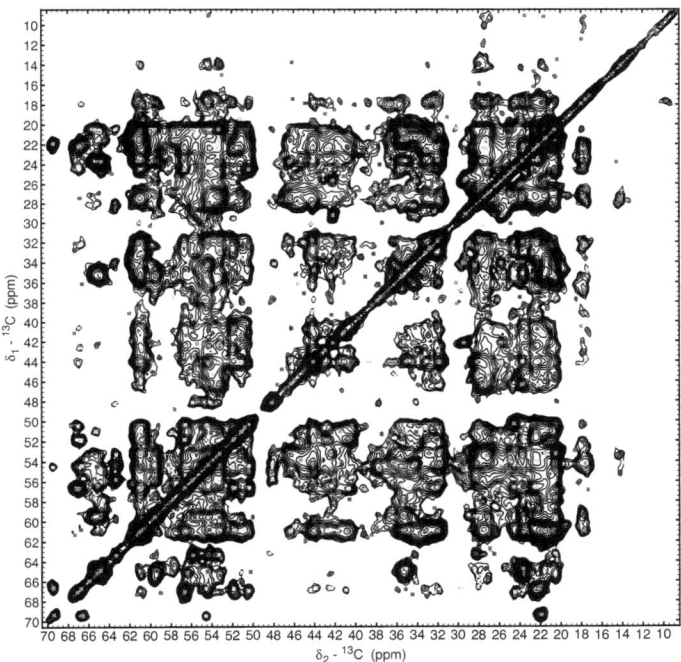

Figure VI.4: PAR Spectrum on homogeneously labeled FgHET-s(218-289) with a mixing time of 16 ms. Peak colors represent the assignment in cycle 7 of the structure calculation: green, intra-residue; yellow, sequential; blue, medium-/ long-range; red, not assigned. The according plots for all other spectra are given in the Appendix, Figs. X.22, X.23, X.24, and X.25

structure calculation protocol described herein, a small number of unassigned resonances is acceptable. This is due to the use of network anchoring and the possibility of peak exclusion from the calculation.

3.4 The Structure of FgHET-s(218-289)

The lowest-energy structure of FgHET-s(218-289) is shown in Fig. VI.5. At first sight, FgHET-s(218-289) seems to resemble HET-s(218-289) (pdb 2KJ3, Van Melckebeke et al. (2010)) but a more careful investigation shows a number of structural differences.

The prolongation of β_{1a} and β_{3a} is most apparent and was already suspected in the previously published model (Wasmer et al. (2010)). As noted before, the 'flexible loop' (region undetectable in CP experiments around residue 250) of FgHET-s(218-289) is less dynamic than in HET-s(218-289). This was apparent in both H/D-exchange and solid-state NMR data (Wasmer et al. (2010)). This change can be explained by the arrangement of residues 258 to 261 in a rigid β-sheet structure and the accompanying shortening of the flexible region before residue 258. The structure is therefore in full agreement with the previously published data for this region.

Another difference between the two fibril structures occurs just at the end of β_{1a} and β_{3a}, within β-arc 1 (Fig. VI.5). The deviation at residues 229 and 230 (265 and 266) forming this β-arc was already apparent from the strong differences in the secondary chemical shifts at these positions (Figure 4(b) in Wasmer et al. (2010)) but the origin of these was speculative. In the structure we obtained for FgHET-s(218-289), this region is well defined and a more detailed comparison was possible. Both ψ_{229} (ψ_{265}) and ϕ_{230} (ϕ_{266}) change by about 180°. This leads to a rotation of 180° of the carbonyl group of E229 (R265) and the attached amide group of F230 (T266) but a net rotation of 360° (or 0°) and therefore only slight changes of the position of the adjacent amino acid side-chains. Most important, the in-/out-order of the side-chains with regard to the hydrophobic core is identical for both arrangements. β-arc 1 is an ab-arc in HET-s(218-289) (see "Standard conformations of two-residue β-arcs" in Hennetin et al. (2006) for definition.) and a bl-arc (or, more exact a V-bl arc), the most common form of two-residue β-arcs (Hennetin et al. (2006)), in FgHET-s(218-289). The secondary chemical shifts of the backbone atoms that are very sensitive to the dihedral angles are consistent with the different arcs 1 in HET-s(218-289) and FgHET-s(218-289) (Wishart and

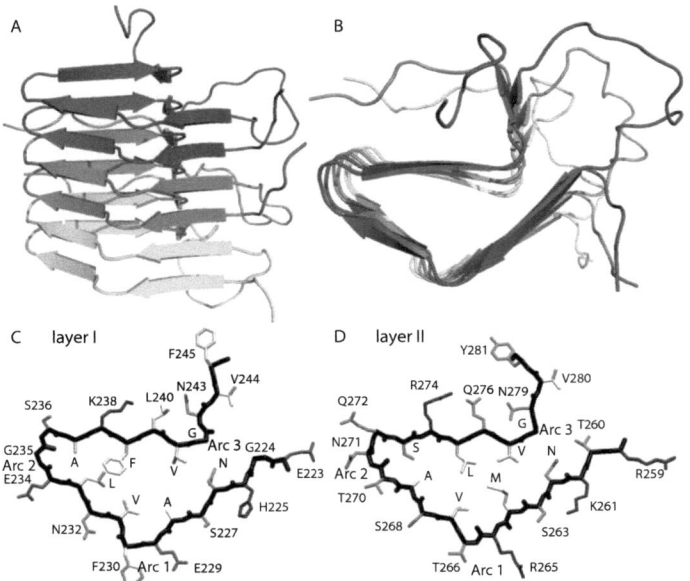

Figure VI.5: Structure of the hydrophobic core of FgHET-s(218-289). **A, B** side and top-view of three FgHET-s(218-289) molecules within the amyloid fibrils in cartoon representation. Each molecule is colored uniquely. The shown arrangement is exactly the same as implemented in the structure calculation. The two layers of the β-solenoid formed by one molecule are shown in **C** (residues 223 to 245) and **D** (residues 259 to 281). Hydrophobic side-chains are colored white, polar ones are green and negatively and positively charged ones are shown in red and blue, respectively. The protein backbone is shown in black; only heavy atoms (N, C, O, S) are shown.

123

Nip (1998)).

The orientation of the 4th β-sheet (β_{2b} and β_{4b}) could not be determined unambiguously. The structure calculation also yielded results with both β_{2b} and β_{4b} rotated about the β-strand direction by 180°, i.e. with the side-chain order inverted (Fig. VI.5). Exactly the same behavior was observed for HET-s(218-289) in the initial calculations (Wasmer et al. (2008a)) but the problem could be resolved by using additional ambiguous distance restraints later (Van Melckebeke et al. (2010)). As a direct consequence it also remains unclear if the 3rd β-arc of FgHET-s(218-289) is a 2-residue or a 3-residue β-arc. The somewhat special position of β_{2b} and β_{4b}, which are not directly involved in the formation of the main hydrophobic core results in a much lower number of possible structural contacts with the well-ordered parts and are likely the cause for the structural underdetermination of this region. However, this also means that the structure of the main hydrophobic core formed by the three remaining β-sheets is not even slightly dependent on the orientation of β_{2b} / β_{4b}.

The C-terminus, residues after Y281, are very poorly defined in the structure calculation. There are some restraints between the aromatic residues in this region and the structurally well-defined core that however do not fix these residues as it is the case in the newest structure of HET-s(218-289) (pdb 2KJ3, Van Melckebeke et al. (2010)). If this part of the protein is indeed disordered or if it is only not well determined by the observed restraints remains to be seen.

4 Conclusions and Discussion

4.1 The Structure Calculation

The structure of HET-s(218-289) had been calculated using a total of 5 different NMR samples and 9 solid-state NMR spectra (Wasmer et al. (2008a)). In an extensive evaluation of the structure calculation procedure it turned out that not all spectra are required to obtain the correct fold (Van Melckebeke et al. (2010)). It was however shown, that spectra from dilute and mixed samples are essential to obtain the correct fold as unambiguously intra- or intermolecular restraints can be derived from them. The restraints from homogeneously labeled samples on the other hand lead to a better defined fold in general.

In the approach we described herein, we used the minimal set of 3 samples, i.e. dilute, mixed and homogeneous. The overall amounts of samples used are 13 mg

U-[^{13}C, ^{15}N], 5 mg U-^{13}C, 5 mg U-^{15}N and 7 mg unlabeled FgHET-s(218-289) that can be cheaply produced as ^{15}N ammonium chloride and ^{13}C glucose are the sole isotopically labeled compounds needed. Additionally, the reduced number of 5 spectra were recorded in less than 20 days of measurement time on an 850 MHz solid-state NMR spectrometer using a standard 3.2 mm triple-resonance probe. This time could again be significantly reduced by NMR probes with increased sensitivity or further optimized sample preparation techniques.

Finally, the structure calculation process has now been implemented in a fully automated fashion using CYANA 3.0d and needs about 1 day for a calculation of all 7 cycles on a cluster with at least 100 CPUs. We optimized only 2 parameters to account for the lower quality of the solid-state NMR data when compared to solution-NMR NOESY spectra. We expect these to be the optimal values for any structure calculation from data of similar quality and the structure calculation outcome is very robust with regard to small changes of these parameters. The only complementary data still required to obtain the correct fibrillar fold of FgHET-s(218-289) were the H/D-exchange rates (Wasmer et al. (2010)) from which we could derive the position of hydrogen-bonds involving amide protons.

The protocol we developed may in principle be applied to any protein fibril forming single filaments. However, it may easily be adapted for fibrils of more complex architecture, e.g. with multiple molecules per asymmetric unit cell or wit multiple identical molecules forming one layer of the fibrils. We could show, that structure determination of amyloid fibrils by solid-state NMR is possible even for samples that produce NMR spectra of lower quality and with and inferior resonance distribution compared to HET-s(218-289). Additionally, we significantly reduced the amount and costs of protein samples and also of NMR measurement time. With the approach at hand, structure determination of amyloid fibrils will become feasible for many systems currently under investigation.

4.2 The Structure

The structure of the amyloid fibrils formed by FgHET-s(218-289), a homologue of HET-s(218-289), was determined at atomic resolution. The comparison of the fibrils formed by each of these two proteins reveals an intriguingly high structural similarity, as both exhibit a triangular hydrophobic core, stabilized by 31-37 H-bonds per molecule, with exactly the same order of the amino acid side-chains—though the amino acid types are of course not identical. Such a fold of FgHET-s(218-289) has been correctly anticipated before by a homology model of HET-s(218-289) (Wasmer et al. (2010)).

With the high-resolution structure available now, also slight differences between the structures can be examined (Fig. VI.6). The only apparent difference in the core region occurs at β-arc 1 (between residues 229 and 230, 265 and 266), that is of a different type in the two fibrils, which however does not largely influence the overall shape of the hydrophobic core. Due to the different primary structure, however, HET-s(218-289) has a higher number of charged, water-accessible sidechains (Van Melckebeke et al. (2011); Wasmer et al. (2010)) on the outside of the core that can form salt-bridges (Lange et al. (2009)). These may well be the reason for the higher stability of HET-s(218-289) with regard to chemical denaturation by the salt GuHCl ($m_{1/2}$ = 5.4 M, while $m_{1/2}$ = 3.0 M for FgHET-s(218-289) (Wasmer et al. (2010))).

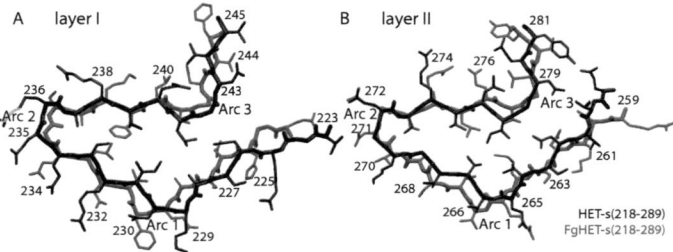

Figure VI.6: Structures of the hydrophobic cores of FgHET-s(218-289) (red) and HET-s(218-289) (black). A first (residues 223 to 245) and B second layer (residues 259 to 281) of one molecule within the β-solenoid. The molecules were manually aligned on residues 232-240 and 268-276 for the two layers, respectively. Note the structural differences in arc 1 (ab-arc in HET-s, V-bl arc in FgHET-s) and arc 3.

The prolongation of the β-sheet composed of $β_{1a}$ and $β_{3a}$ towards the N-terminus represents a more articulate difference. It is probably stabilized by the two oppositely charged residues E223 and R259 but is less stable, which is evidenced by the faster H/D-exchange rates in this region as compared to the core (Wasmer et al. (2010)).

HET-s(218-289) and FgHET-s(218-289) do not form entropically favorable fibrils consisting of stoichometrically mixed molecules (our unpublished data and personal communication with S. J. Saupe). With the structural similarity observed, this seems counterintuitive, but such a fibril would exhibit numerous unfavorable close encounters of charged side-chains of the same polarity. However, these interactions do not prevent the highly efficient seeding of HET-s(218-289) with fibrils of FgHET-s(218-289) and vice versa (Wasmer et al. (2010)).

Apart from the small differences discussed above, the fold of HET-s(218-289) (Wasmer et al. (2008a)) is highly preserved in FgHET-s(218-289) although the native hosts of the two proteins, *P. anserina* and *F. graminearum*, have an estimated evolutionary divergence time of 400 MYrs. (Wasmer et al. (2010)), which is readily apparent in the diverged sequences. The function of FgHET-s, however, is unknown and that of HET-s is highly speculative (Saupe (2007)). Taking into account the high degree of structural conservation, a specific function of this protein fold seems likely, but is currently unknown.

VI The Structure of FgHET-s(218-289)

Conclusions and Outlook

In the thesis at hand, state-of-the-art high-resolution solid-state NMR techniques were employed to structurally characterize amyloid-forming proteins, with an emphasis on prions in their infectious state. We could show that structure determination of amyloid fibrils at atomic resolution is feasible by solid-state NMR. Our approach requires a set of isotopically labeled protein samples (about 30 mg in total) and some complementary structural information, ideally the H/D-exchange rates of the amide protons. The exact prerequisites depend largely on the achievable quality of the NMR spectra; the structure of HET-s(218-289), for example, can also be derived solely from solid-state NMR data of high quality without the need for any complementary information (Van Melckebeke et al. (2010)).

In addition, we could demonstrate that solid-state NMR is an invaluable tool to structurally characterize and compare amyloid fibrils in general. Even for samples of a quality too poor for an atomic-resolution structure determination, i.e. if the sample exhibits a considerable degree of disorder or polymorphism, valuable structural information can be drawn from the solid-state NMR spectra.

The structure-determination protocol developed within this work may be extended to different fibrillar arrangements and is in principle applicable to any structure repetitive in 1 dimension (1-dimensional crystal symmetry). This condition is by definition fulfilled by all protein fibrils that include disease-related proteins as Aβ from Alzheimer's disease, α-synuclein from Parkinson's disease, and the mammalian prion involved in Creutzfeld-Jakob. However, an additional level of complexity may be introduced for fibrils exhibiting polymorphism and/or containing several molecules per asymmetric unit cell. These issues may render a structure calculation more difficult, however, in principle the described method may be used to determine the structures of these proteins in their disease-related aggregated states. This will lead to a better understanding of their roles in the course of the respective disease. The structural knowledge may finally allow for the directed development of prospective drugs targeting the process of fibril formation in the context of these highly prevalent neurodegenerative diseases.

Additionally the large and until now structurally poorly characterized class of natively aggregated proteins, including in particular functional amyloids (Greenwald and Riek (2010)), is accessible by our approach. A very interesting class of such proteins is for example constituted by a number of hormones, which have been shown to be stored in an amyloid conformation by Maji et al. (2009).

Chapter X

Appendix

X Appendix

1 Non-infectious pH 3 Fibrils

Comparison of pH 3 and pH 7 fibrils. The full carbonyl and aliphatic regions of PDSD spectra of pH 3 and pH 7 fibrils are shown in Figs. X.1 and X.2, respectively. The N–C HETCOR spectra of these two samples are given in Fig. X.3.

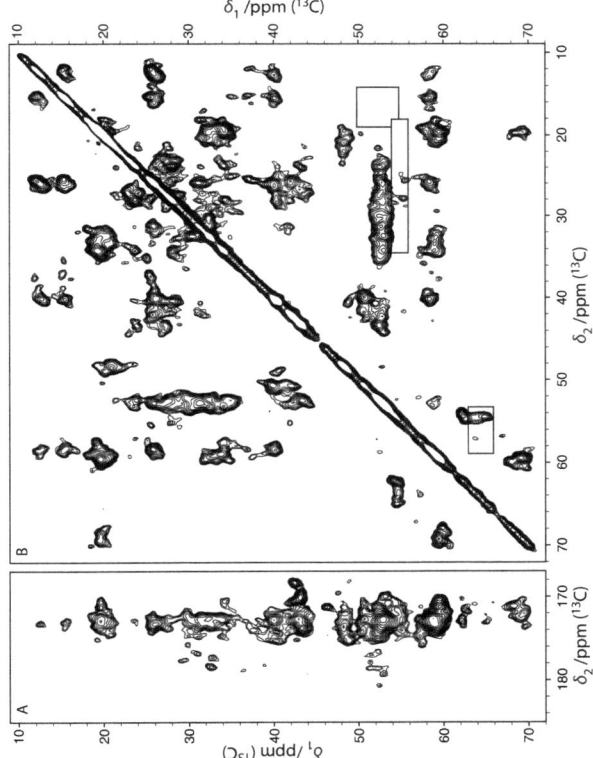

Figure X.1: PDSD spectrum of pH 3 fibrils. **A, B** carbonyl and aliphatic regions of HET-s(218-289) fibrils formed at pH 3. The black boxes enclose peaks discussed in the main text; the same areas here as in the spectrum of pH 7 fibrils (Suppl. Fig. X.2). The spectrum was recorded at 13 kHz MAS with a mixing period of 50 ms and 90 kHz SPINAL-64 ^{1}H-decoupling during t_1 and t_2.

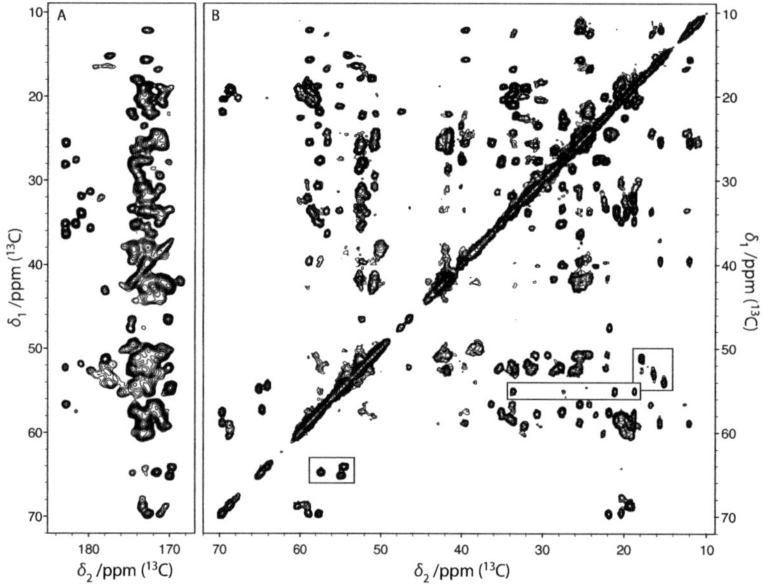

X Appendix

Figure X.2: PDSD spectrum of pH 7 fibrils. **A**, **B** carbonyl and aliphatic regions of HET-s(218-289) fibrils formed at pH 7. The black boxes enclose peaks from residues discussed in the main text (A237, A247, A248, V264, S227, S263 and S273); the same areas here as in Fig. X.1. The spectrum was recorded at 10 kHz MAS with a mixing period of 50 ms and 90 kHz SPINAL-64 ^{1}H-decoupling during t_1 and t_2.

134

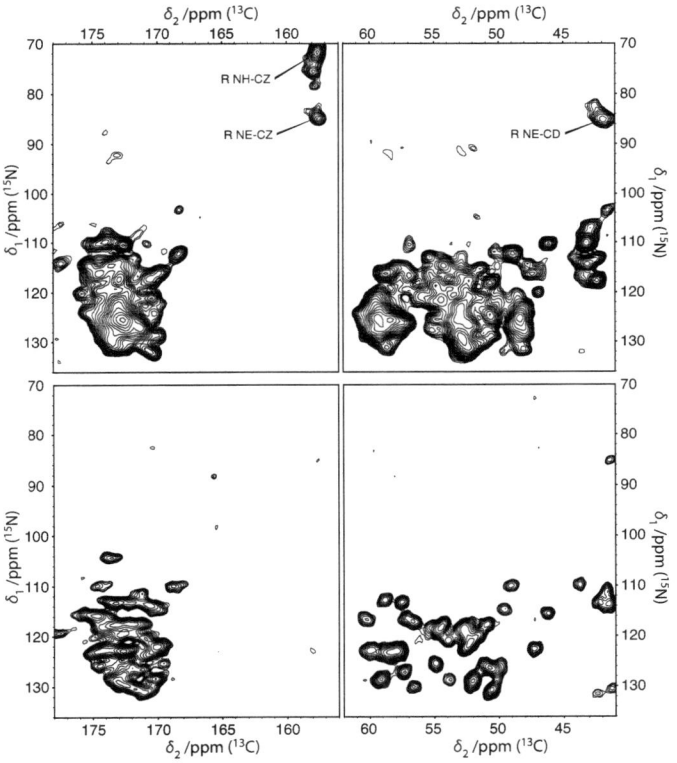

Figure X.3: N–C HETCOR spectra of pH 3 fibrils (blue, upper half) and of pH 7 fibrils (red, lower half). The spectra were recorded at 13 kHz and 10 kHz MAS, respectively with a mixing period of 3 ms and 90 kHz SPINAL-64 ^{1}H-decoupling during t_1 and t_2.

X Appendix

Assignment of spin systems in the rigid parts. For the assignment of the spin systems, a CP-TOBSY spectrum was recorded. Fig. X.4 shows the spectrum with assignments obtained from concurrent analysis of TOBSY, PDSD and the N–C HETCOR spectrum recorded for this sample. A comprehensive list of the assignments for the rigid parts of the pH 3 fibrils is given in Suppl. Table X.1.

	C'	C^α	C^β	$C^{\gamma(1/2)}$	$C^{\delta(1/2)}$	$C^{\epsilon(1)/\zeta}$	N^ϵ	$N^{\zeta/\eta}$
A_1	-	48.4	22.2					
A_2	-	48.4	20.8					
A_3	-	48.2	19.1					
D/N	-	50.4	40.0	175.2				
E_1	-	53.3	30.4	32.5	178.5			
E_2	-	-	-	33.7	180.7			
G_1	170.0	43.0						
G_2	168.3	42.4						
H	-	-	27.2	129.4	118.0	134.6	-	
I_1	-	58.4	37.0	25.5/ 15.2	11.9			
I_2	-	58.3	40.1	25.9/15.4	12.7			
K	-	53.2	34.1	23.7	28.0	40.3		-
L	-	-	-	27.3	21.4/24.0			
R	-	52.8	31.5	25.8	42.0	157.6	85.0	72.0
S_1	-	54.8	64.1					
S_2	-	54.7	62.5					
T_1	-	59.4	69.3	19.7				
T_2	-	59.8	67.9	-				
V_1	-	59.0	33.7	-				
V_2	-	59.4	32.6	-				
V_3	-	57.2	34.7	-				
V_4	-	59.9	31.7	18.2				

Table X.1: Chemical shifts in ppm of spin systems assigned in the CP-TOBSY, PDSD and N–C HETCOR spectra (Figs. X.1, X.3 and X.4) probing the rigid parts, referenced to TMS. A "-" marks atoms for which no chemical shift assignment was possible because of resonance overlap or missing signals.

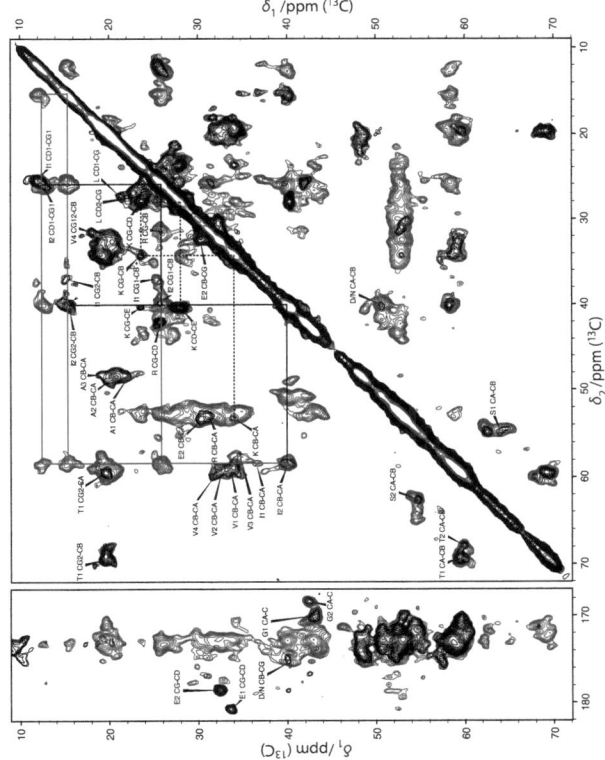

Figure X.4: ^{13}C–^{13}C-TOBSY (black) and 50 ms PDSD (blue) spectrum of pH 3 fibrils. The assignments were obtained by analysis of TOBSY, PDSD and N–C HETCOR spectra and are numbered arbitrarily. The spectra shown were recorded at 13 kHz MAS with 90 kHz SPINAL-64 ^{1}H-decoupling during t_1 and t_2. The mixing times were 5 ms and 50 ms for the TOBSY and the PDSD spectrum, respectively.

2 Full-length HET-s

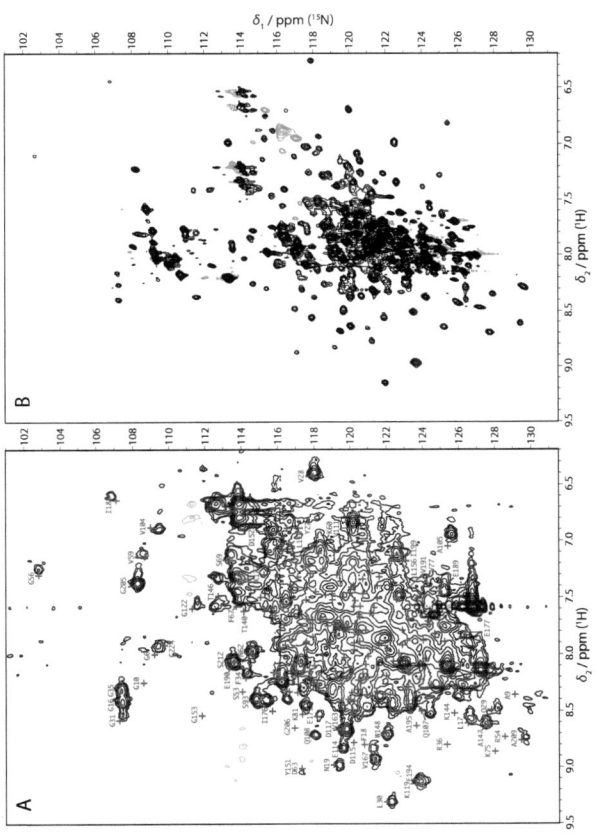

Figure X.5: $^1H-^{15}N$ HSQC TROSY correlation experiments recorded on **A** the HET-s sample used in this study and **B** U-[^{2}H, ^{15}N] HET-s, produced in the same way, both after the complete purification but before fibrillation. The red crosses in **A** indicate peaks expected for soluble HET-s(1-227) (solution NMR experiments for HET-s(1-227) resonance assignment not shown). For the isolated expected peaks, the assignment is given. The experiments were recorded on a Bruker Avance700 spectrometer with a cryo-probe, by Carolin Buhtz and Roland Riek.

X Appendix

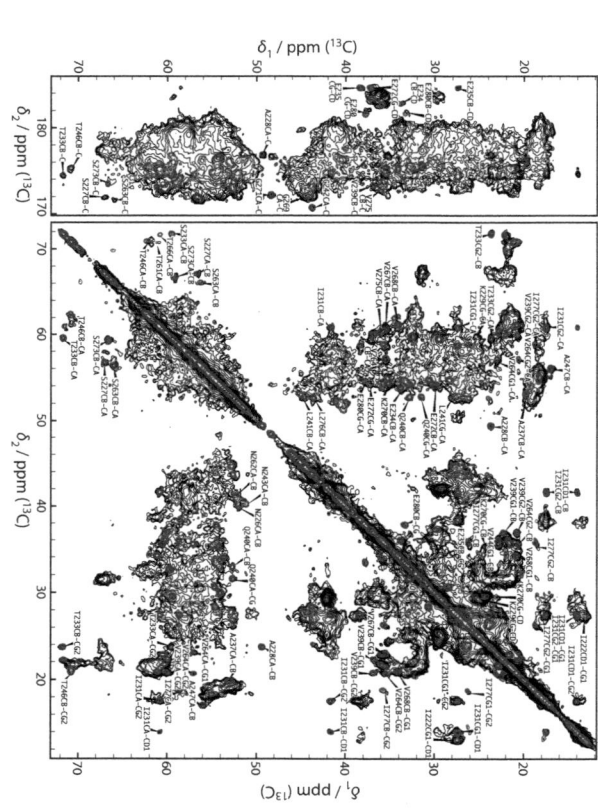

Figure X.6: 100 ms C–C DARR (Takegoshi et al. (2001)) spectra of HET-s(218-289) (blue) and HET-s (black). Peaks that appear for other samples and that are at least partly resolved are labeled using the known resonance assignments of HET-s(218-289) (Wasmer et al. (2008a)).

140

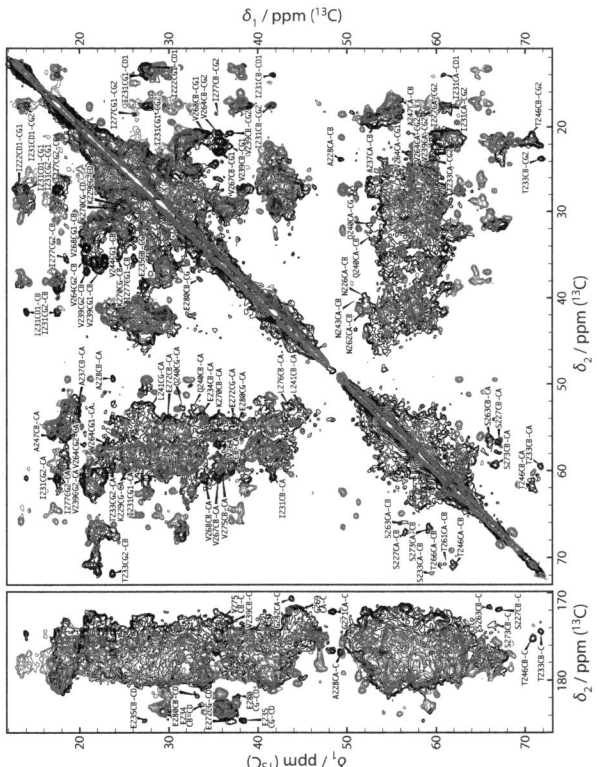

Figure X.7: 100 ms C–C DARR (Takegoshi et al. (2001)) spectra of HET-s(1-227) (red) and HET-s (black). Peaks that appear for other samples and that are at least partly resolved are labeled using the known resonance assignments of HET-s(218-289) (Wasmer et al. (2008a)).

3 HET-s(218-289) Inclusion Bodies

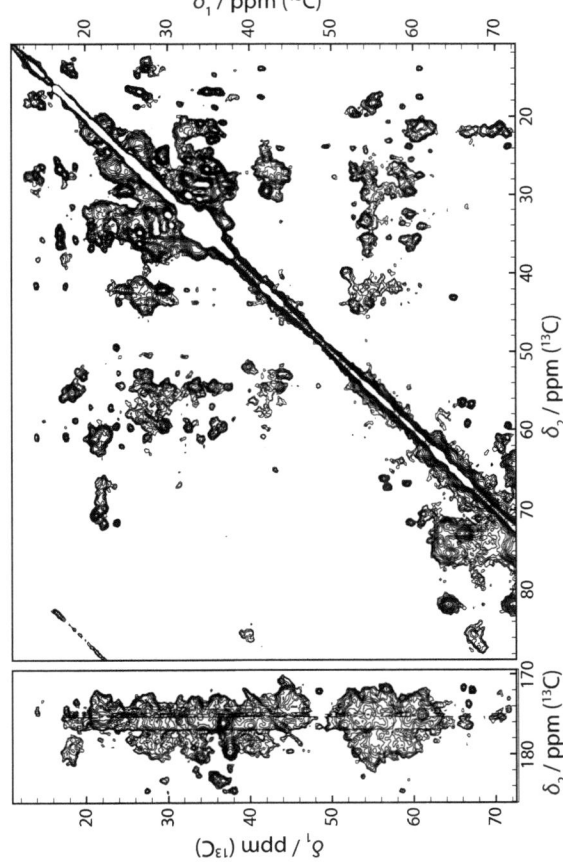

Figure X.8: Carbonyl and aliphatic regions of a $^{13}C-^{13}C$ solid-state NMR correlation spectrum (PDSD with a mixing time of 50 ms) of raw HET-s(218-289) IBs

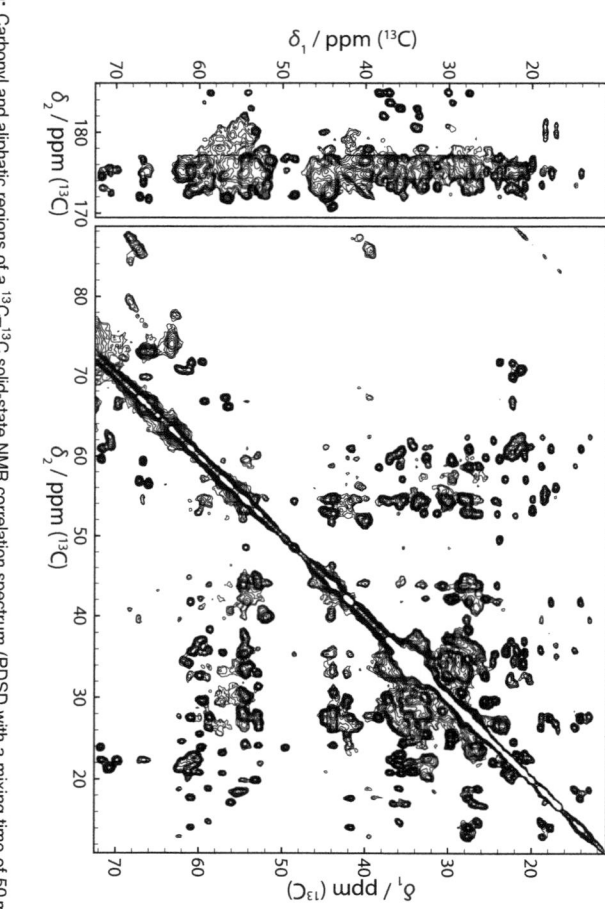

Figure X.9: Carbonyl and aliphatic regions of a ^{13}C–^{13}C solid-state NMR correlation spectrum (PDSD with a mixing time of 50 ms) of purified HET-s(218-289) IBs

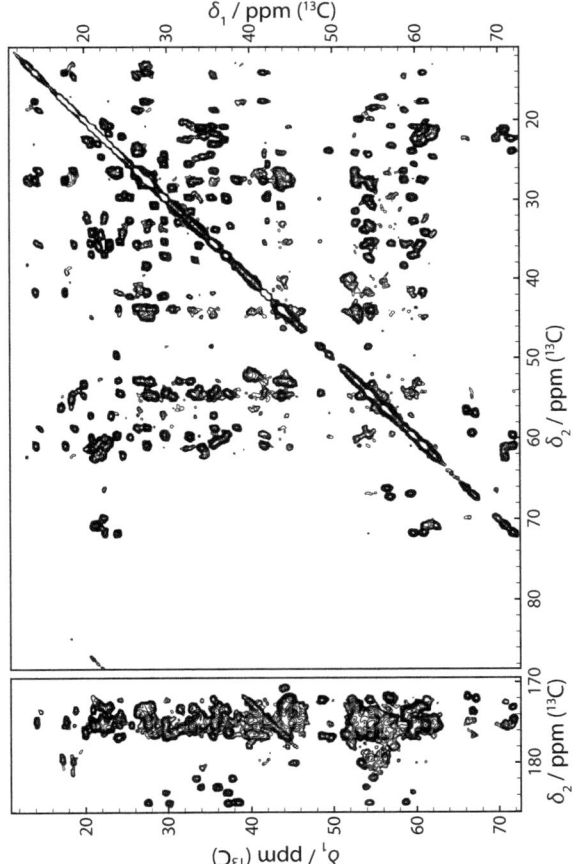

Figure X.10: Carbonyl and aliphatic regions of a ^{13}C–^{13}C solid-state NMR correlation spectrum (PDSD with a mixing time of 50 ms) of HET-s(218-289) IBs *in vitro* amyloid fibrils

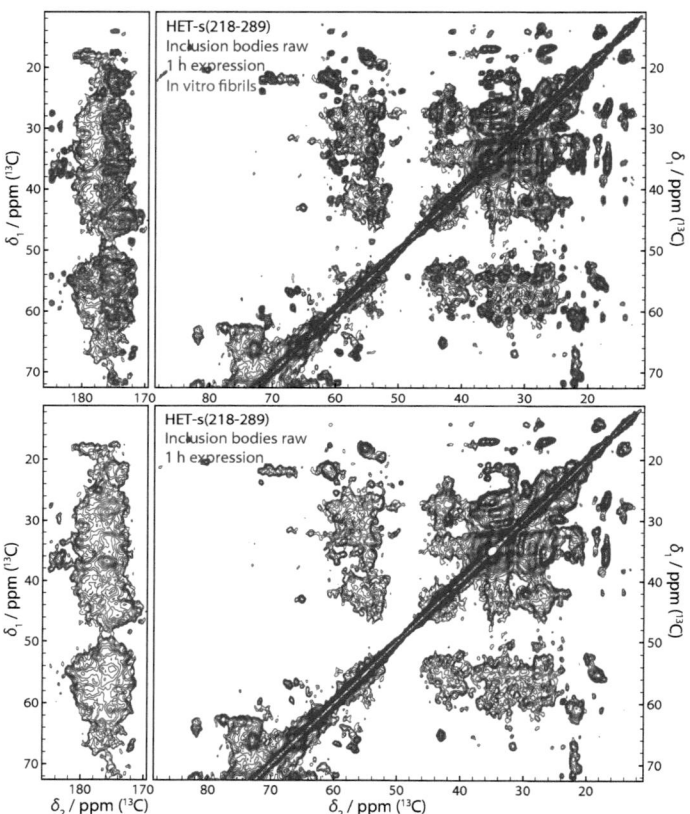

Figure X.11: Carbonyl and aliphatic regions of ^{13}C–^{13}C solid-state NMR correlation spectra (PDSD with a mixing time of 50 ms) of raw HET-s(218-289) IBs obtained from cells 1 h after induction of the recombinant expression (blue contours) and *in vitro* fibrillized HET-s(218-289) (red contours). Most of the signals assigned for the purified fibrils can be observed in the IB Spectrum, although they are very weak due to the small amount of HET-s(218-289) obtained in this short expression period.

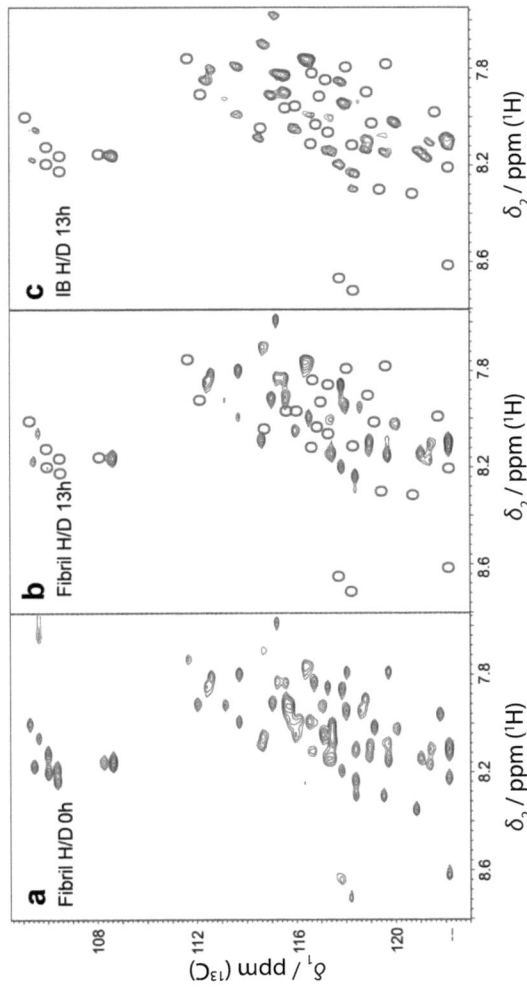

Figure X.12: H/D-exchange on HET-s(218-289) amyloid fibrils and purified IBs. Fast HMQC spectra in d_6-DMSO containing 0.1 % d_1-TFA of **a** fully protonated fibrils, **b** partially hydrogen exchanged fibrils and **c** partially hydrogen exchanged purified inclusion bodies. The red circles mark positions of peaks that are not observable anymore after the exchange period, indicating relatively high exchange rates for the corresponding residues. Fibrils and inclusion bodies were H/D-exchanged for 13 h at 4 °C before measurement.

4 FgHET-s(218-289)

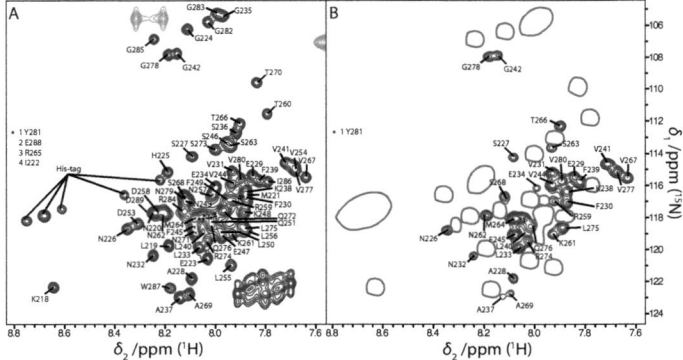

Figure X.13: HSQC spectra for H/D-exchange of **A** fully protonated FgHET-s(218-289) and **B** after incubation in deuterated buffer for 4 weeks.

	N	C'	C^α	C^β	$C^\gamma/C^{\gamma 1}/$ $C^{\gamma 2}$	$C^\delta/C^{\delta 1}/$ $C^{\delta 2}$	$C^\epsilon/C^{\epsilon 1}/$ $C^{\epsilon 2}/C^{\epsilon 3}$	$C^\zeta/C^{\zeta 2}/$ $C^{\zeta 3}$	$C^{\eta 2}$	$N^{\delta 2}/N^\epsilon/$ $N^{\epsilon 2}$	$N^\zeta/N^{\eta 1}$
E223		175.3	54.0	30.1	36.6	183.5					
G224	108.2	172.6	45.1								
H225	118.6			31.4	128.7	119.9	135.2				
N226	126.5	174.0	51.8	40.6		178.1				114.4	
S227	117.1	172.7	56.5	67.2							
A228	122.2	176.2	50.3	24.4							
E229	122.1	175.2	54.4	34.7	36.0	183.3					
F230	119.2	173.4	59.4	35.1	140.4	131.6	130.0				
V231	123.2	174.8	60.9	35.5	20.4/21.6						
N232	128.3	173.0	53.0	42.2		177.0					
L233	125.1	176.2	62.5	45.4	28.2	25.1					
E234	121.9	174.8	54.0	34.0	36.2	184.1					
G235	109.0	173.7	48.0								
S236	122.7	175.3	55.7	63.3							
A237	125.7	175.4	53.1	20.4							
K238	118.1	174.7	54.4	37.8	26.2	30.1	42.5				33.4
F239	124.6	172.7	54.2	44.1	139.6	130.8	128.5				
L240	130.6	174.0	53.4	46.3	27.7	23.7					
V241	129.9	173.3	61.1	31.9	22.1/23.6						
G242	111.8	172.1	44.1								
N243	114.6	174.8	51.0	41.3							
V244	118.0	177.5	60.8	34.7	21.4/23.6						
F245	125.4										
S246	120.2		57.7	64.5							
D258	124.6	175.1	52.8	42.8	178.4						
R259	125.5	176.3	54.3	31.6	28.0	44.1		159.5		85.2	
T260	122.6	174.6	66.5	69.4	22.0						
K261	129.2	173.9	54.7	33.9	25.6	29.3	42.1				33.4
N262	125.5	175.4	51.9	42.7		174.1				112.7	
S263	117.3	172.8	56.6	66.8							
M264	122.2	175.2	54.4	39.0	32.6						
R265	126.8	174.5	55.1	34.7	28.2	43.9		159.8		85.3	72.9
T266	118.1	171.7	65.3	64.7	23.6						
V267	124.0	173.6	57.7	35.2	20.4/21.5						
S268	123.7	172.7	56.2	65.1							
A269	128.1	175.3	51.1	22.4							
T270	113.1	175.2	57.7	69.5	19.9						
N271	119.0	172.4	55.0	37.6		177.8				111.9	
Q272	122.0	175.9	54.3	28.2	34.0	180.9				111.4	
S273	117.4	174.2	59.6	65.7							
R274	118.4	173.8	54.9	36.4	27.9	44.3		159.6		85.1	72.2
L275	120.6	173.8	54.1	44.7	27.5						
Q276	120.7	174.0	52.9	32.8	31.6	176.2				104.3	
V277	128.5	174.0	60.6	33.0	22.7/21.6						
G278	113.8	169.7	44.0								
N279	109.9	176.1	50.7	41.3		177.5				115.0	
V280	120.9	174.1	61.4	33.6	21.2						
Y281	128.4			38.6							
W287	120.8		55.9	29.6	111.7	128.5	140.1/ 120.1	115.2/ 120.7	123.4		

Table X.2: Extent of chemical shift assignment of FgHET-s(218-289). All chemical shifts are given in ppm and were referenced to DSS.

X Appendix

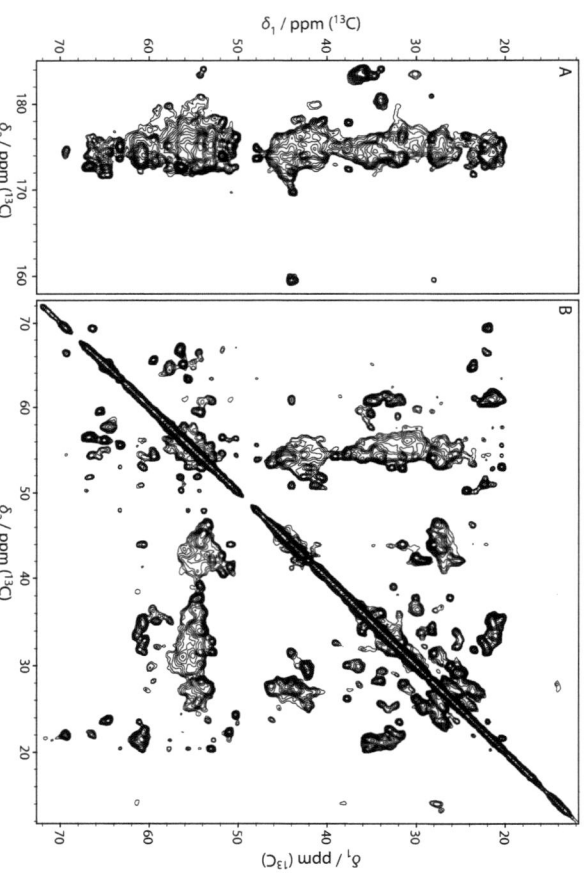

Figure X.14: **A** Carbonyl and **B** aliphatic regions of the $^{13}C-^{13}C$ DARR spectrum of FgHET-s(218-289) with a mixing time of 100 ms. In **B**, the same region as in Fig. V.5 in the main text is displayed.

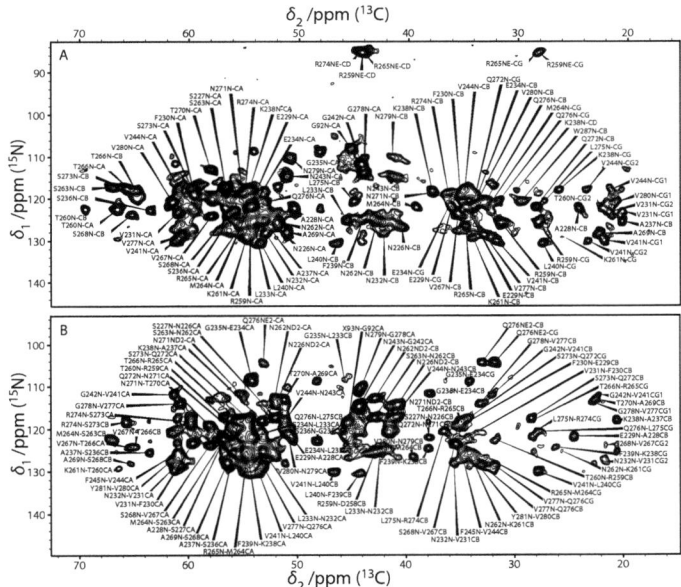

Figure X.15: Aliphatic regions of **A** the 2D N(CA)CX and **B** the 2D N(CO)CX spectra (Detken et al. (2001)) with labels for all peaks that are both visible and expected. Note that almost all unlabeled peaks can be explained by sequential transfers (from residue i to $i \pm 1$) during the homonuclear mixing period. Both spectra use a 50 ms DARR / MIRROR period for the C–C mixing.

X Appendix

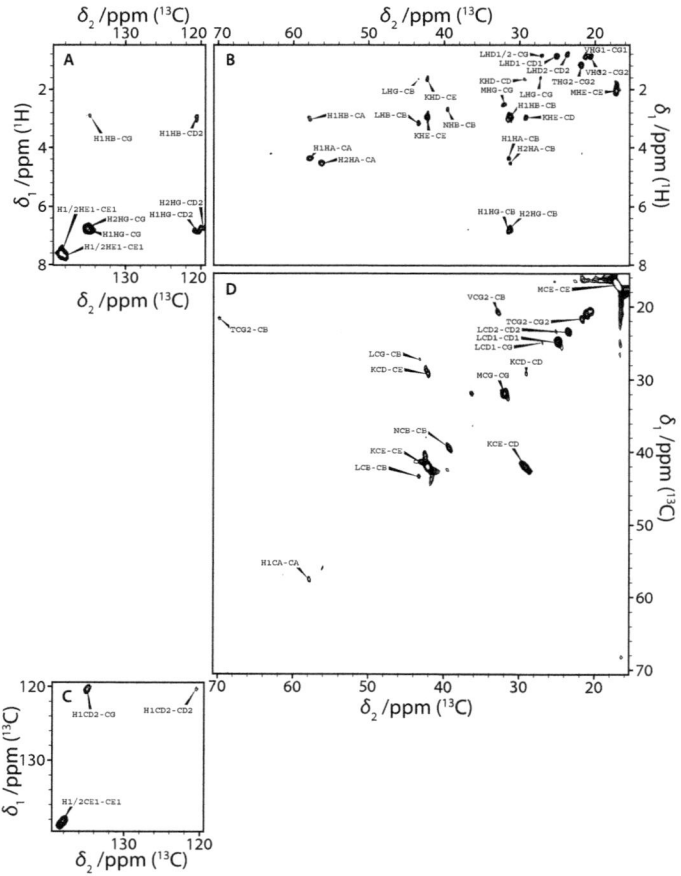

Figure X.16: Aromatic and aliphatic regions of the INEPT-H–(C)–C (**A, B**) and the INEPT-(H)–C–C spectrum (**C, D**) sensitive to flexible residues of U-[^{13}C, ^{15}N] FgHET-s(218-289) fibrils (Andronesi et al. (2005); Siemer et al. (2006a)). The unambiguously assignable peaks that lead to the identification of the amino acid types of the flexibly disordered residues are labeled. The homonuclear transfer was achieved by a 5 ms TOBSY (Hardy et al. (2001)) mixing period for both spectra.

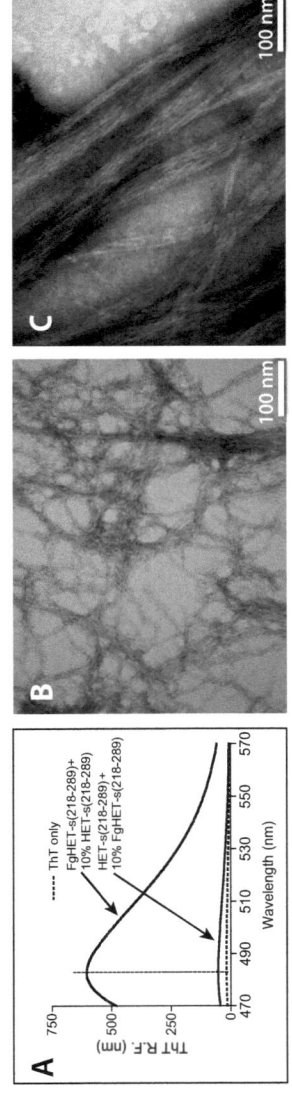

Figure X.17: FgHET-s(218-289) and HET-s(218-289) fibril morphology in cross-seeding assays. **A** Cross-seeding assays wherein FgHET-s(218-289) fibrils, formed in the presence of 10 % preformed HET-s(218-289) fibrils, induce ThT fluorescence while HET-s(218-289), formed in the presence of 10 % preformed FgHET-s(218-289) fibrils, show only a residual ThT signal. Note that the same results have been obtained for non cross-seeded fibrils (see Fig. V.2B in Chapter V). The excitation wavelength was 450 nm and emission was recorded from 470 nm to 570 nm. ThT and protein concentrations of 25 µM and 10 µM respectively were used. Electronic micrographs of **B** FgHET-s(218-289) and **C** HET-s(218-289) fibrils obtained from cross-seeding assays. Note that the appearance of cross-seeded fibrils is very similar to non-seeded assays of FgHET-s(218-289) and HET-s(218-289) (see Fig. V.2A), respectively.

X   Appendix

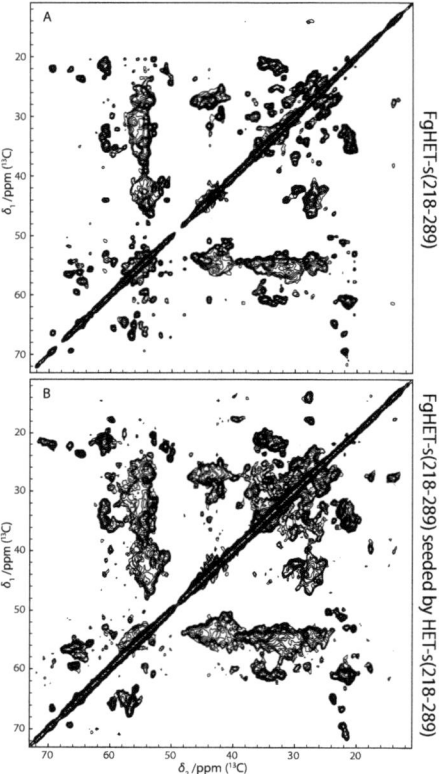

Figure X.18: Aliphatic regions of a PDSD spectrum of U-[^{13}C, ^{15}N] labeled samples of **A** FgHET-s(218-289) and **B** FgHET-s(218-289) seeded by sonicated HET-s(218-289) fibrils with 100 ms DARR mixing. For this short mixing times short-range (intra-residue and sequential) correlations are dominant. Note that the overall signal intensity is lower in spectrum **B**, which causes low-intensity peaks (e.g. C^α-C^α correlations) to fall below the noise level. However, the chemical shifts of the observed peaks in both spectra are identical, indicating an identical fold of the two samples. The cross-seeded spectra show broader lines and signs of polymorphism. However, these have been observed in some unseeded preparations. Both aliphatic and carbonyl regions of the DARR spectrum of FgHET-s(218-289) are given in Fig. X.14.

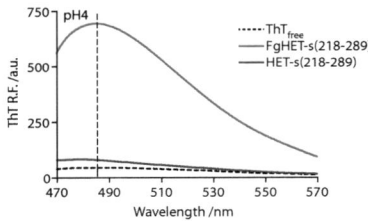
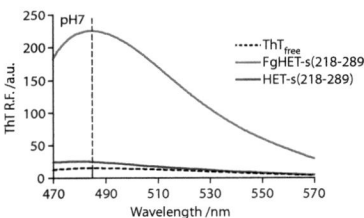

Figure X.19: FgHET-s(218-289) fibrils induce ThT-fluorescence while HET-s(218-289) fibrils do not at both pH 4 and pH 7.

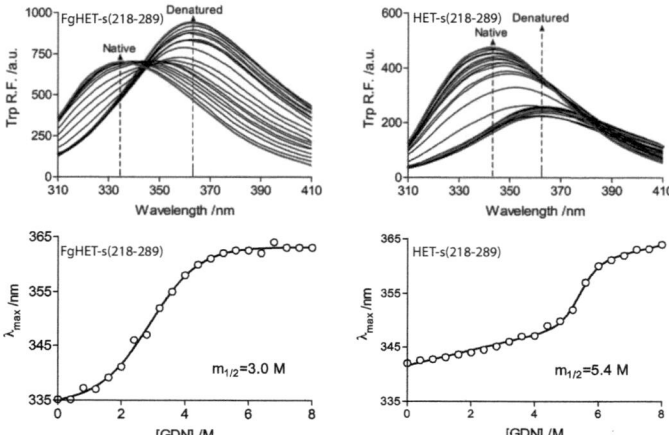

Figure X.20: FgHET-s(218-289) vs. HET-s(218-289) fibrils stability against chemical denaturation by GuHCl (GDN). The two upper panes show UV absorption spectra of samples at different concentrations of GuHCl (blue: 0 M, red: 8 M). In the lower panes λ_{max}, the wavelength at which maximum absorption occurs is plotted against the GuHCl concentration. $m_{1/2}$ gives the concentration, at which λ_{max} is halfway between the native and the fully denatured states.

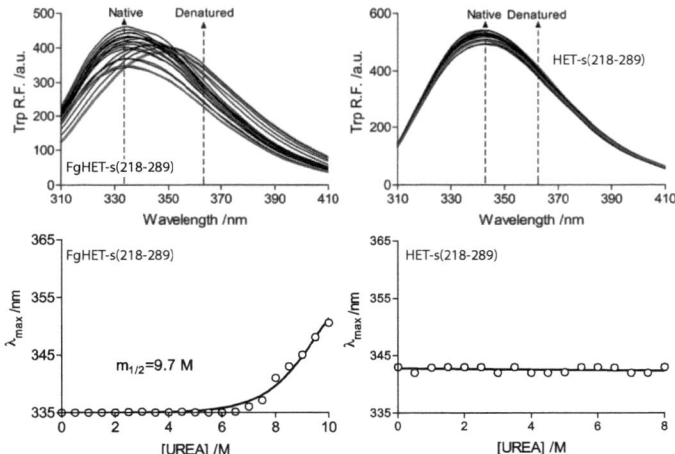

Figure X.21: FgHET-s(218-289) vs. HET-s(218-289) fibrils stability against chemical denaturation by Urea. The two upper panes show UV absorption spectra of samples at different concentrations of Urea. In the lower panes λ_{max}, the wavelength at which maximum absorption occurs is plotted against the Urea concentration. $m_{1/2}$ gives the concentration, at which λ_{max} is halfway between the native and the fully denatured states.

X Appendix

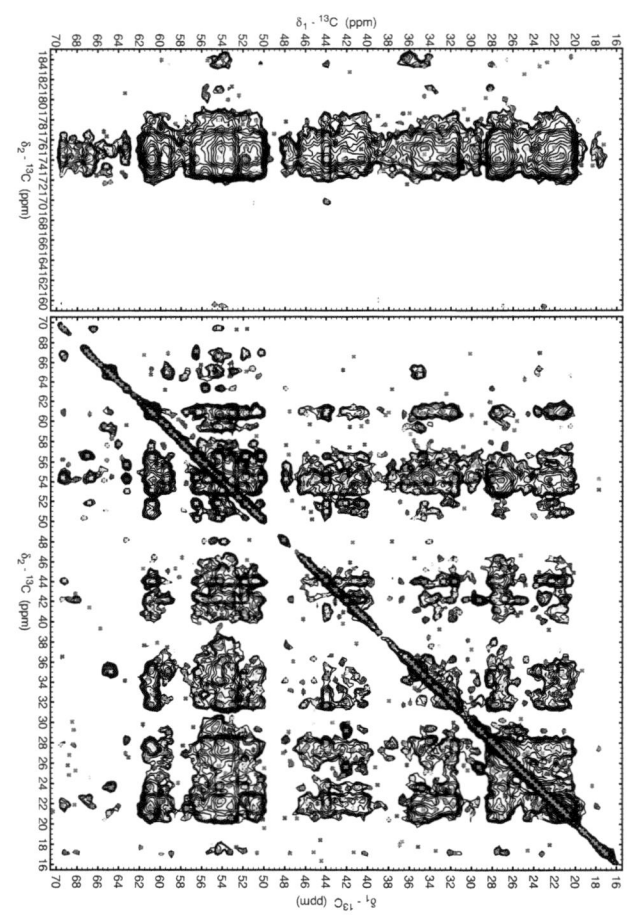

Figure X.22: PAR Spectrum on homogeneously labeled FgHET-s with a mixing time of 8 ms. Peak colors represent the assignment in cycle 7 of the structure calculation: green, intra-residue; yellow, sequential; blue, medium-/long-range; red, not assigned.

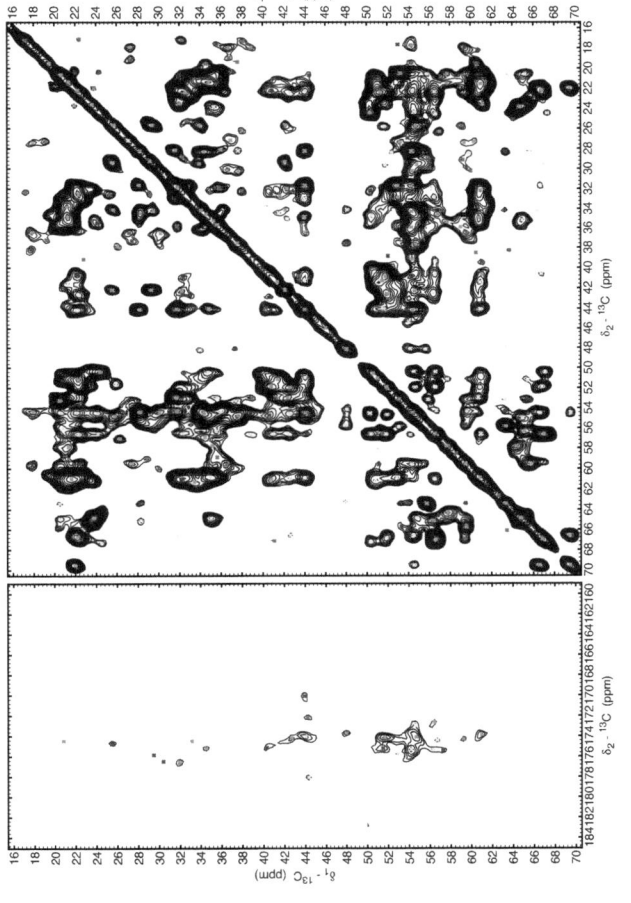

Figure X.23: PAR Spectrum on dilutely labeled FgHET-s with a mixing time of 10 ms. Peak colors represent the assignment in cycle 7 of the structure calculation: green, intra-residue; yellow, sequential; blue, medium-/ long-range; red, not assigned.

X Appendix

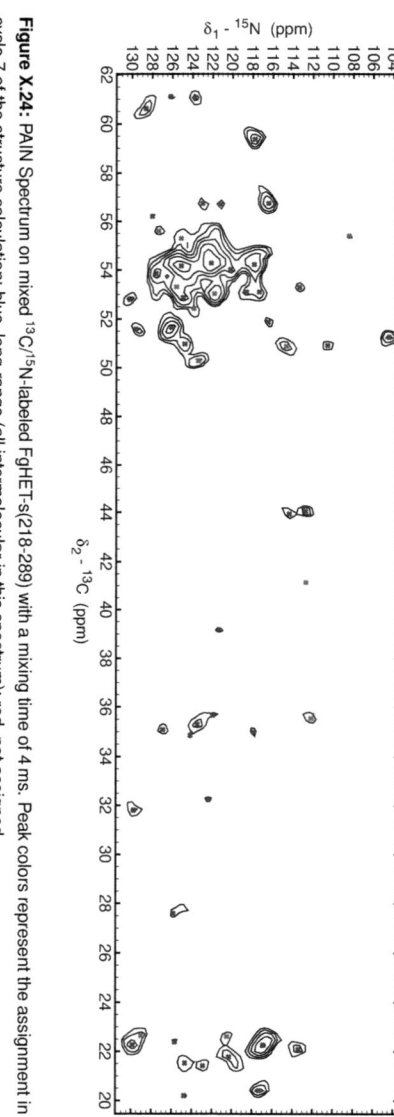

Figure X.24: PAIN Spectrum on mixed ^{13}C/^{15}N-labeled FgHET-s(218-289) with a mixing time of 4 ms. Peak colors represent the assignment in cycle 7 of the structure calculation: blue, long-range (all intermolecular in this spectrum); red, not assigned.

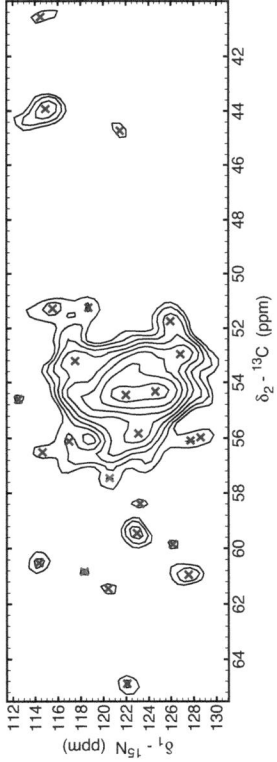

Figure X.25: NHHC Spectrum on mixed ^{13}C/^{15}N-labeled FgHET-s(218-289) with a mixing time of 250 µs. Peak colors represent the assignments: blue, long-range (all intermolecular in this spectrum); red, not assigned.

X Appendix

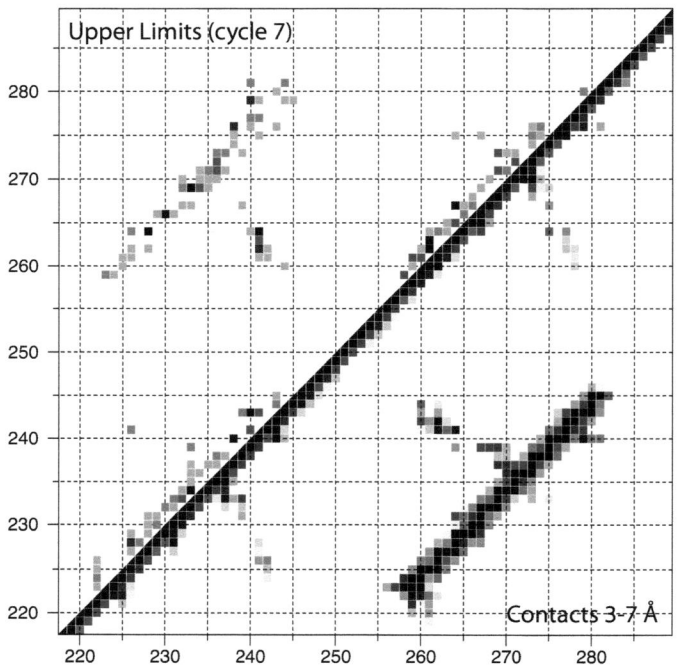

Figure X.26: Residue-residue plot of contacts observed in the NMR spectra (upper left, black: 5 contacts) and distances < 7 Å in the final structure (light grey: 6-7 Å, black: ≤ 3 Å).

5 Sample preparation protocol

established December 6, 2006
last updated on February 22, 2011

for the preparation of
HET-s(1-289), HET-s(218-289) pH 7 fibrils,
FgHET-s(218-289) pH 7 fibrils and HET-s(218-289) pH 3 fibrils

General Remarks

- This protocol is not extensive at all; use common sense to fill the gaps.

- Keep the fractions mentioned to be discarded for analysis on an SDS Tricine gel until you are sure that everything worked.

- Always separate s'natant and pellet immediately after a centrifugation step to avoid re-suspension of pellets

- Don't trust this protocol too much - it surely contains several errors!

5.1 Recombinant Expression in *E. coli*

Transformation – into chemically competent E.coli BL21(DE3); always work sterile.

1. Thaw competent cells 10' @ 4 °C. Thaw DNA-solution.

2. Add 2 µl (or 1 µg) of DNA to 10 µl competent cells. Swirl pipet through solution to mix. Do *not* vortex!

3. Keep on ice for 30'.

4. Heat shock cells for 1'30" at 43 °C. Do not shake.

5. Add 90 µl of room temperature LB.

6. Incubate at 37 °C, 200 rpm for 1 h. Pre-warm an LB agar plate containing the appropriate antibiotic (usually Kanamycin for HET-s(218-289) and FgHET-s(218-289) and Ampicillin for HET-s(1-289)).

7. Dispense 50 µl - 100 µl on the agar plate.

8. Incubate at 37 °C for 12 h to 18 h. If the transformation was successful, 100 to 1000 colonies should grow.

9. Immediately start the first culture from the plate before storing it at 4 °C.

Expression – always work sterile (except last step).

1. Freshly transform vector into competent E.coli BL21(DE3) host cells. Do not use colonies from old agar plates or glycerol stock cells.

2. If you want to express isotopically labeled Het-s, prepare the appropriate M9 medium (Recipe appended as last page of this protocol)

3. Inoculate 2 (or more) pre-cultures (6 ml each) of desired medium (LB or M9, pre-warmed to 37 °C) with a colony from the plate. Incubate at 37 °C, 200 rpm.

4. Measure OD_{600} as soon as pre-culture becomes turbid. Always dilute culture before OD_{600} reaches 1.
   The expected doubling times are about 30' in LB and 60' in M9 containing glucose. For M9 using glycerol and $NaHCO_3$ as sole carbon sources, doubling will occur after 75'.

5. Inoculate pre-warmed main cultures with an arbitrary amount taken from *one* of the pre-culture s. Discard pre-culture s. Incubate main cultures at 37°C, 220 rpm.

6. Measure OD_{600} as soon as cultures become turbid. Induce with 1 mM IPTG at $OD_{600} = 0.5$ for M9 or 1.0 for LB medium.

7. Harvest cells by successive centrifugation in a 1000 ml centrifugation tube (10' @ 18'000 g) after about five hours (or more for HET-s(218-289) as it forms protease resistant inclusion bodies and won't be digested). Immediately freeze the pellet at $-80°C$. Discard s'natant.

5.2 Purification

Inclusion Body Extraction – always keep on ice and centrifuge at 4°C unless indicated otherwise!

1. Thaw cell pellet on ice. Re-suspend in 30 ml of buffer **A** and transfer to a small centrifugation tube.
2. Microfuidize once on ice.
3. Centrifuge for 1 h @ 15'000 g, discard s'natant.
4. Re-suspend pellet in 20 ml buffer **A** + 2% **Triton X-100**.
5. Incubate 20' at 37°C, 230 rpm.
6. Centrifuge 10' @ 50'000 g, discard s'natant.
7. Re-suspend pellet in 25 ml buffer **A** (without Triton)
8. Incubate 2 h @ 37°C, 230 rpm.
9. Centrifuge 10' @ 50'000 g, discard s'natant.
10. Re-suspend pellet in 25 ml **B**. This works best if you first thoroughly re-suspend the pellet in < 10 ml 0.5 M NaCl, 50 mM Tris·HCl and then add dry GuHCl until saturation. *The proper re-suspension might be difficult and is a crucial step!*
11. Incubate at 60°C overnight and sonicate the sample.
12. Centrifuge $\geq$ 1 h @ 250'000 g, 25°C; discard pellet.

You may skip steps 4 to 9 (I now usually do so), i.e. directly resuspend the first pellet in **B**.

Ni-Affinity Batch Binding – this is my preferred method for HET-s purification now, as column clogging cannot occur. Usually done in a 50 ml falcon tube.
For HET-s(1-289) elute by using buffer **C** with 8 M Urea instead of GuHCl (Most probably this won't make a difference).

1. Wash ≈ 5 ml of Ni-Sepharose 6 FastFlow resin 3× in ≥ 20 ml buffer **B** (Centrifuge 10' @ 500 g, remove s'natant with pipet)

2. Add the s'natant from the previous step to the resin and incubate in the rotator for 30'.

3. Centrifuge 10' @ 500 g, remove s'natant with pipet.

4. Repeat the two previous steps if you have more sample to load; as often as necessary.

5. Wash the resin at least once in buffer **B**.

6. Put the resin into a column (don't forget to put a frit first).

7. Wash this column with at least 5× your resin volume.

8. Wait until no buffer is left above the resin, close column outlet; you may put a second frit on top now.

9. Add ≤ 10 ml buffer **C** and open the outlet. Start collecting fractions – I usually take 1 ml; do not use larger fractions if you want a high conc. in the end

10. Collect until around 20 ml (at least if you do this for the first time)

11. Wash the column with 20 % EtOH and store at RT – the resin can be used a lot of times and even be recharged with Ni (see manual).

The amount of protein in each fraction can be easily estimated by UV absorption measurements. For HET-s(218-289) the protein concentration is given by $c = 1.24 * OD_{280}$ mg/ml.

If you want to prepare a mixed or diluted sample, now is the time to pool the appropriate fractions from the different runs.

Desalting – only use nano-pure water and degassed and 0.2 μm-filtered buffer. *Never* inject air onto the column! If you want to desalt small amounts (≤ 1 ml) with a high final concentration consider using a 5 ml desalting column (values in brackets) instead of the HiPrep 26/10.

1. Fill fraction collector with 15 ml Falcon tubes.
2. Connect the 10 ml sample loop (1 ml) to the Äkta chromatographic system.
3. Only HET-s(1-289) : Add DTT to a final concentration of 20 mM to your purified sample.
4. Connect inlet A1 to buffer **D** (**D "pH 2"** if you want to prepare pH3 fibrils) for HET-s(218-289) or (**A** + 1 % DTT) for HET-s(1-289) and wash the loop until UV and conductivity are stable — "Autozero".
5. Install the HiPrep 26/10 column (or HiTrap Desalting 5 ml for ≤ 1 ml of sample).
6. Wash column with 265 ml (25 ml) of buffer, until UV and conductivity are stable.
7. Set injection valve to "load" and load up to 10 ml (1 ml) sample.
8. Choose "Run stored method", select "15" ("11") and run.
9. As soon as "method complete" displays, you can repeat step 7. Your desalted protein will be collected in one or more Falcon tubes.
10. Wash system, column and loop with 20 % Ethanol; store column at room temperature.

5.3 Fibrillation

Fibrillation of FgHET-s(218-289) and HET-s(218-289) pH 7 fibrils

1. Pool the fractions you want to fibrillize from the desalting runs into a Falcon tube.

2. Measure absorption at 280 nm and calculate the protein concentration by $c = 1.24\,\text{mg} \cdot A_{280}$ (assuming pure, unfolded Het-s(218-298)).

3. Insert pH electrode into the solution and set to constant readout.

4. Add 3 M Tris until pH reaches about 7.4 (you will need to add about 8 % of the total volume).

5. Wait and look how your fibrils form. Let stand for 1 day. Stirring or shaking does not influence the quality of the sample (as far as I can tell).

6. Wash in pure water several times, centrifuge down at 10'000 g for a few minutes.

Fibrillation of HET-s(218-289) pH 3 fibrils

1. Pool the fractions you want to fibrillize from the desalting runs into a Falcon tube.

2. Measure absorption at 280 nm and calculate the protein concentration by $c = 1.24\,\text{mg} \cdot A_{280}$ (assuming pure, unfolded Het-s(218-298)). Start with a concentration $\geq 1.5\,\text{mg/ml}$.

3. Set pH to 3.00-3.05.

4. Measure OD_{280} and dilute sample to a final concentration of 1 mg/ml.

5. Incubate at 37°C, fast shaking or rotations for 2 days.

6. Wash 2-3× in pure water.

X   Appendix

Remarks for HET-s(1-289) pH 7 Fibrillation   – The first two steps are already given in the section "Desalting".

1. Add Dithiothreitol (DTT) to a concentration of 20 mM prior to desalting.
2. Exchange buffer to 150 mM NaCl, 50 mM Tris HCl pH 8, 1 mM DTT on a HiTrap Desalting 5 ml column (GE Healthcare).
3. The sample will refold and finally fibrillize spontaneously in this buffer. IMPORTANT: Do not centrifuge this sample in a regular falcon tube. From my experience, it will stick to the tube and never re-suspend again.
4. You may remove any remaining salt by dialyzing against pure water.
5. Centrifuge into an MAS rotor at 200,000 g.

Drying   – Partly drying the samples is not necessary and not recommended when using an ultracentrifuge (> 200'000 g) to fill NMR rotors.

1. Wash the fibrils prior to drying in nano-pure water by about 4 centrifugation (10'000 g for about 5') and re-suspension steps.
2. Puncture the lid of the falcon tube with a 0.9 mm needle. (1 to 5 holes will cause evaporation of 6.5 mg/h to 25 mg/h of water.)
3. Set the minimum pressure p to 95 mbar, Δp to 10 mbar and the temperature to 25°C. Let dry N_2 flow by or use silica gel.
4. Dry to a protein concentration of 30 % (Higher values may work – to be determined). The consistence should now be "about that of Roquefort" (A. Siemer, 2006)

This protocol was established following Balguerie et al. (2003) and Ritter et al. (2005) and with the kind help of Christiane Ritter.

Samples for Lange and Meier (2008); Lange et al. (2009); Van Melckebeke et al. (2010, 2011); Wasmer et al. (2008a,b, 2009a,b, 2010, 2011) were prepared according to the procedures described herein.

5.4 Buffers and Media

Buffers for inclusion body extraction and purification (recommended amounts in brackets).

- **A** (500 ml)
    - 150 mM NaCl
    - 5 % Tris-HCl (1 M, pH 8)
- **A + 2 % Triton X-100**   (250 ml)
- **B** (2 l)
    - 7.5 M GuHCl
    - 20 mM Imidazole
    - 0.5 M NaCl
    - 5 % Tris-HCl (1 M, pH 8)
- **C** (250 ml)
    - 7.5 M GuHCl
    - 500 mM Imidazole
    - 0.5 M NaCl
    - 5 % Tris-HCl (1 M, pH 8)
- **D** (2 l)
    - 175 mM Acetic Acid (10 g/l)

The chromatography buffers **B**, **C** and **D** need to be filtered (0.2 µm) and degassed. Store buffers **A**, **B** and **C** at 4 °C. The GuHCl in Buffers **B** and **C** will crystallize out (sometimes) when stored below room temperature. Warm them up at least 4 h before usage.

1 l minimal medium (M9)

ingredient *essential*	amount	stock soln. labelling					
$Na_2HPO_4 \times 7H_2O$ or	12.8 g	50 ml					
$Na_2HPO_4 \times 2H_2O$ or	8.5 g	20x M9					
Na_2HPO_4	6.8 g	salts					
KH_2PO_4	3.0 g						
NaCl	0.5 g		○	○	○	○	○
NH_4Cl	1.0 g	^{15}N?	○	○	○	○	○
Water	add to 0.9 l		(add before $MgSO_4$!)				
$MgSO_4$ (1 M)	1 ml		○	○	○	○	○
$ZnCl_2$ (10 mM)	1 ml	1 ml					
$FeCl_3$ (1 mM)	1 ml	Q 1000					
$CaCl_2$ (100 mM)	1 ml		○	○	○	○	○
Vitamin-Mix (1:100)	10 ml		○	○	○	○	○
Glucose or *Glycerol*	2.5 g	^{13}C?					
or 20% stock soln.	10 ml	^{13}C?	○	○	○	○	○
$NaH^{13}CO_3$ (only for 2-^{13}C)	2.0 g	^{13}C?	○	○	○	○	○
$NaH^{12}CO_3$ (only for 1,3-^{13}C)	2.0 g	^{13}C?	○	○	○	○	○
Water	add to 1 l						
pH	adjust to 7.0		(not really necessary)				
antibiotic							
Kanamyc. 30 mg/ml	1 ml		○	○	○	○	○
Ampicillin 150 mg/ml	1 ml		○	○	○	○	○

sterile filtrate with 0.22 μm filter
store at 4 °C

Bibliography

Bibliography

Andronesi, O. C., Becker, S., Seidel, K., Heise, H., Young, H. S., and Baldus, M. (2005). Determination of membrane protein structure and dynamics by magic-angle-spinning solid-state nmr spectroscopy. *J Am Chem Soc*, 127(37):12965–12974.

Baldus, M., Geurts, D. G., Hediger, S., and Meier, B. H. (1996). Efficient 15n-13c polarization transfer by adiabatic-passage hartmann-hahn cross polarization. *J Magn Reson, Ser A*, 118(1):140–144.

Baldus, M. and Meier, B. H. (1996). Total correlation spectroscopy in the solid state. the use of scalar couplings to determine the through-bond connectivity. *J Magn Reson, Ser A*, 121(1):65–69.

Balguerie, A., Dos Reis, S., Coulary-Salin, B., Chaignepain, S., Sabourin, M., Schmitter, J.-M., and Saupe, S. J. (2004). The sequences appended to the amyloid core region of the het-s prion protein determine higher-order aggregate organization in vivo. *J Cell Sci*, 117(Pt 12):2599–2610.

Balguerie, A., Dos Reis, S., Ritter, C., Chaignepain, S., Coulary-Salin, B., Forge, V., Bathany, K., Lascu, I., Schmitter, J.-M., Riek, R., and Saupe, S. J. (2003). Domain organization and structure-function relationship of the het-s prion protein of podospora anserina. *EMBO J*, 22(9):2071–2081.

Baneyx, F. (1999). Recombinant protein expression in escherichia coli. *Curr Opin Biotechnol*, 10(5):411–421.

Baneyx, F. and Mujacic, M. (2004). Recombinant protein folding and misfolding in escherichia coli. *Nat Biotechnol*, 22(11):1399–1408.

Baxa, U., Cassese, T., Kajava, A. V., and Steven, A. C. (2006). Structure, function, and amyloidogenesis of fungal prions: filament polymorphism and prion variants. *Adv Protein Chem*, 73:125–180.

Baxa, U., Cheng, N., Winkler, D. C., Chiu, T. K., Davies, D. R., Sharma, D., Inouye, H., Kirschner, D. A., Wickner, R. B., and Steven, A. C. (2005). Filaments of the ure2p prion protein have a cross-beta core structure. *J Struct Biol*, 150(2):170–179.

Benkemoun, L., Sabate, R., Malato, L., Dos Reis, S., Dalstra, H., Saupe, S. J., and Maddelein, M.-L. (2006). Methods for the in vivo and in vitro analysis of [het-s] prion infectivity. *Methods*, 39(1):61–67.

Bennett, A., Rienstra, C., Auger, M., Lakshmi, K., and Griffin, R. (1995). Heteronuclear decoupling in rotating solids. *Journal of Chemical Physics*, 103(16):6951–6958.

Böckmann, A., Gardiennet, C., Verel, R., Hunkeler, A., Loquet, A., Pintacuda, G., Emsley, L., Meier, B. H., and Lesage, A. (2009). Characterization of different water pools in solid-state nmr protein samples. *J Biomol NMR*, 45(3):319–327.

Bousset, L., Thomson, N. H., Radford, S. E., and Melki, R. (2002). The yeast prion ure2p retains its native alpha-helical conformation upon assembly into protein fibrils in vitro. *EMBO J*, 21(12):2903–2911.

Bibliography

Brünger, A. T., Adams, P. D., Clore, G. M., DeLano, W. L., Gros, P., Grosse-Kunstleve, R. W., Jiang, J. S., Kuszewski, J., Nilges, M., Pannu, N. S., Read, R. J., Rice, L. M., Simonson, T., and Warren, G. L. (1998). Crystallography & nmr system: A new software suite for macromolecular structure determination. *Acta Crystallogr D Biol Crystallogr*, 54(Pt 5):905–21.

Burum, D. P. and Ernst, R. R. (1980). Net polarization transfer via a j-ordered state for signal enhancement of low-sensitivity nuclei. *J Magn Reson (1969)*, 39(1):163–168.

Carrio, M., Gonzalez-Montalban, N., Vera, A., Villaverde, A., and Ventura, S. (2005). Amyloid-like properties of bacterial inclusion bodies. *J Mol Biol*, 347(5):1025–1037.

Castellani, F., van Rossum, B., Diehl, A., Schubert, M., Rehbein, K., and Oschkinat, H. (2002). Structure of a protein determined by solid-state magic-angle-spinning nmr spectroscopy. *Nature*, 420(6911):98–102.

Chan, J. C. C., Oyler, N. A., Yau, W.-M., and Tycko, R. (2005). Parallel beta-sheets and polar zippers in amyloid fibrils formed by residues 10-39 of the yeast prion protein ure2p. *Biochemistry*, 44(31):10669–10680.

Chen, L., Kaiser, J. M., Polenova, T., Yang, J., Rienstra, C. M., and Mueller, L. J. (2007). Backbone assignments in solid-state proteins using j-based 3d heteronuclear correlation spectroscopy. *J Am Chem Soc*, 129(35):10650–10651.

Chiti, F. and Dobson, C. M. (2006). Protein misfolding, functional amyloid, and human disease. *Annu Rev Biochem*, 75:333–366.

Chothia, C. and Lesk, A. M. (1986). The relation between the divergence of sequence and structure in proteins. *EMBO J*, 5(4):823–826.

Clantin, B., Hodak, H., Willery, E., Locht, C., Jacob-Dubuisson, F., and Villeret, V. (2004). The crystal structure of filamentous hemagglutinin secretion domain and its implications for the two-partner secretion pathway. *Proc Natl Acad Sci U S A*, 101(16):6194–6199.

Collinge, J. (2001). Prion diseases of humans and animals: their causes and molecular basis. *Annu Rev Neurosci*, 24:519–550.

Cornell, W., Cieplak, P., Bayly, C., Gould, I., Merz, K., Ferguson, D., Spellmeyer, D., Fox, T., Caldwell, J., and Kollman, P. (1996). A second generation force field for the simulation of proteins, nucleic acids, and organic molecules (vol 117, pg 5179, 1995). *J Am Chem Soc*, 118(9):2309–2309. Journal of the American Chemical Society.

Cornilescu, G., Delaglio, F., and Bax, A. (1999). Protein backbone angle restraints from searching a database for chemical shift and sequence homology. *J Biomol NMR*, 13(3):289–302.

Coustou, V., Deleu, C., Saupe, S., and Begueret, J. (1997). The protein product of the het-s heterokaryon incompatibility gene of the fungus podospora anserina behaves as a prion analog. *Proc Natl Acad Sci U S A*, 94(18):9773–9778.

Dalstra, H. J. P., van der Zee, R., Swart, K., Hoekstra, R. F., Saupe, S. J., and Debets, A. J. M. (2005). Non-mendelian inheritance of the het-s prion or het-s prion domains determines the het-s spore killing system in podospora anserina. *Fungal Genet Biol*, 42(10):836–47.

De Paepe, G., Bayro, M. J., Lewandowski, J., and Griffin, R. G. (2006). Broadband homonuclear correlation spectroscopy at high magnetic fields and mas frequencies. *J Am Chem Soc*, 128(6):1776–1777.

De Paepe, G., Lewandowski, J. R., Loquet, A., Bockmann, A., and Griffin, R. G. (2008). Proton assisted recoupling and protein structure determination. *J Chem Phys*, 129(24):245101.

Detken, A., Hardy, E., Ernst, M., and Meier, B. (2002). Simple and efficient decoupling in magic-angle spinning solid-state nmr: the xix scheme. *Chemical Physics Letters*, 356(3-4):298–304.

Detken, A., Hardy, E. H., Ernst, M., Kainosho, M., Kawakami, T., Aimoto, S., and Meier, B. H. (2001). Methods for sequential resonance assignment in solid, uniformly 13c, 15n labelled peptides: quantification and application to antamanide. *J Biomol NMR*, 20(3):203–21.

Diaz-Avalos, R., King, C.-Y., Wall, J., Simon, M., and Caspar, D. L. D. (2005). Strain-specific morphologies of yeast prion amyloid fibrils. *Proc Natl Acad Sci U S A*, 102(29):10165–10170.

Diercks, T., Coles, M., and Kessler, H. (1999). An efficient strategy for assignment of cross-peaks in 3d heteronuclear noesy experiments. *J Biomol NMR*, 15(2):177–180.

Dos Reis, S., Coulary-Salin, B., Forge, V., Lascu, I., Begueret, J., and Saupe, S. J. (2002). The het-s prion protein of the filamentous fungus podospora anserina aggregates in vitro into amyloid-like fibrils. *J Biol Chem*, 277(8):5703–5706.

Etzkorn, M., Bockmann, A., Lange, A., and Baldus, M. (2004). Probing molecular interfaces using 2d magic-angle-spinning nmr on protein mixtures with different uniform labeling. *J Am Chem Soc*, 126(45):14746–14751.

Ferguson, N., Becker, J., Tidow, H., Tremmel, S., Sharpe, T. D., Krause, G., Flinders, J., Petrovich, M., Berriman, J., Oschkinat, H., and Fersht, A. R. (2006). General structural motifs of amyloid protofilaments. *Proc Natl Acad Sci U S A*, 103(44):16248–16253.

Fowler, D. M., Koulov, A. V., Balch, W. E., and Kelly, J. W. (2007). Functional amyloid–from bacteria to humans. *Trends Biochem Sci*, 32(5):217–224.

Fung, B. M., Khitrin, A. K., and Ermolaev, K. (2000). An improved broadband decoupling sequence for liquid crystals and solids. *J Magn Reson*, 142(1):97–101.

Ganapathy, S., van Gammeren, A. J., Hulsbergen, F. B., and de Groot, H. J. M. (2007). Probing secondary, tertiary, and quaternary structure along with protein-cofactor interactions for a helical transmembrane protein complex through 1h spin diffusion with mas nmr spectroscopy. *J Am Chem Soc*, 129(6):1504–1505.

Georgiou, G. and Valax, P. (1999). Isolating inclusion bodies from bacteria. *Methods Enzymol*, 309:48–58.

Ginalski, K. (2006). Comparative modeling for protein structure prediction. *Curr Opin Struct Biol*, 16(2):172–177.

Goldman, M. and Jacquinot, J. (1982). Nuclear-spin diffusion in a rare spin species. *Journal De Physique*, 43(7):1049–1058.

Govaerts, C., Wille, H., Prusiner, S. B., and Cohen, F. E. (2004). Evidence for assembly of prions with left-handed beta-helices into trimers. *Proc Natl Acad Sci U S A*, 101(22):8342–8347.

Greenwald, J., Buhtz, C., Ritter, C., Kwiatkowski, W., Choe, S., Maddelein, M.-L., Ness, F., Cescau, S., Soragni, A., Leitz, D., Saupe, S. J., and Riek, R. (2010). The mechanism of prion inhibition by het-s. *Mol Cell*, 38(6):889–899.

Greenwald, J. and Riek, R. (2010). Biology of amyloid: Structure, function, and regulation. *Structure*, 18(10):1244–60.

Grissmer, S., Nguyen, A. N., Aiyar, J., Hanson, D. C., Mather, R. J., Gutman, G. A., Karmilowicz, M. J., Auperin, D. D., and Chandy, K. G. (1994). Pharmacological characterization of five cloned voltage-gated k+ channels, types kv1.1, 1.2, 1.3, 1.5, and 3.1, stably expressed in mammalian cell lines. *Mol Pharmacol*, 45(6):1227–1234.

Grommek, A., Meier, B. H., and Ernst, M. (2006). Distance information from proton-driven spin diffusion under mas. *Chem Phys Lett*, 427(4-6):404–409.

Guntert, P., Dotsch, V., Wider, G., and Wuthrich, K. (1992). Processing of multi-dimensional nmr data with the new software prosa. *Journal of Biomolecular NMR*, 2(6):619–629.

Guntert, P., Mumenthaler, C., and Wuthrich, K. (1997). Torsion angle dynamics for nmr structure calculation with the new program dyana. *J Mol Biol*, 273(1):283–298.

Hardy, E. H., Detken, A., and Meier, B. H. (2003). Fast-mas total through-bond correlation spectroscopy using adiabatic pulses. *J Magn Reson*, 165(2):208–218.

Hardy, E. H., Verel, R., and Meier, B. H. (2001). Fast mas total through-bond correlation spectroscopy. *J Magn Reson*, 148(2):459–464.

Hartmann, S. and Hahn, E. (1962). Nuclear double resonance in the rotating frame. *Physical Review*, 128(5):2042–2053.

Hediger, S., Meier, B. H., and Ernst, R. R. (1995). Adiabatic passage hartmann-hahn cross polarization in nmr under magic angle sample spinning. *Chem Phys Lett*, 240(5-6):449–456.

Hediger, S., Meier, B. H., Kurur, N. D., Bodenhausen, G., and Ernst, R. R. (1994). Nmr cross polarization by adiabatic passage through the hartmann–hahn condition (aphh). *Chem Phys Lett*, 223(4):283–288.

Heise, H., Hoyer, W., Becker, S., Andronesi, O. C., Riedel, D., and Baldus, M. (2005). Molecular-level secondary structure, polymorphism, and dynamics of full-length alpha-synuclein fibrils studied by solid-state nmr. *Proc Natl Acad Sci U S A*, 102(44):15871–15876.

Helmus, J. J., Surewicz, K., Nadaud, P. S., Surewicz, W. K., and Jaroniec, C. P. (2008). Molecular conformation and dynamics of the y145stop variant of human prion protein in amyloid fibrils. *Proc Natl Acad Sci U S A*, 105(17):6284–6289.

Helmus, J. J., Surewicz, K., Surewicz, W. K., and Jaroniec, C. P. (2010). Conformational flexibility of y145stop human prion protein amyloid fibrils probed by solid-state nuclear magnetic resonance spectroscopy. *J Am Chem Soc*, 132(7):2393–403.

Hennetin, J., Jullian, B., Steven, A. C., and Kajava, A. V. (2006). Standard conformations of beta-arches in beta-solenoid proteins. *J Mol Biol*, 358(4):1094–1105.

Hong, M. (1999). Resonance assignment of 13c/15n labeled solid proteins by two- and three-dimensional magic-angle-spinning nmr. *J Biomol NMR*, 15(1):1–14.

Hornemann, S., Korth, C., Oesch, B., Riek, R., Wider, G., Wüthrich, K., and Glockshuber, R. (1997). Recombinant full-length murine prion protein, mprp(23-231): purification and spectroscopic characterization. *FEBS Lett*, 413(2):277–81.

Hoshino, M., Katou, H., Hagihara, Y., Hasegawa, K., Naiki, H., and Goto, Y. (2002). Mapping the core of the beta(2)-microglobulin amyloid fibril by h/d exchange. *Nat Struct Biol*, 9(5):332–336.

Jaroniec, C. P., MacPhee, C. E., Bajaj, V. S., McMahon, M. T., Dobson, C. M., and Griffin, R. G. (2004). High-resolution molecular structure of a peptide in an amyloid fibril determined by magic angle spinning nmr spectroscopy. *Proc Natl Acad Sci U S A*, 101(3):711–716.

Kajava, A. V., Squire, J. M., and Parry, D. A. D. (2006). Beta-structures in fibrous proteins. *Adv Protein Chem*, 73:1–15.

Kajava, A. V. and Steven, A. C. (2006). Beta-rolls, beta-helices, and other beta-solenoid proteins. *Adv Protein Chem*, 73:55–96.

Keller, R. (2004). *The Computer Aided Resonance Assignment Tutorial.* CANTINA Verlag.

Kishimoto, A., Hasegawa, K., Suzuki, H., Taguchi, H., Namba, K., and Yoshida, M. (2004). beta-helix is a likely core structure of yeast prion sup35 amyloid fibers. *Biochem Biophys Res Commun*, 315(3):739–745.

Koditz, J., Ulbrich-Hofmann, R., and Arnold, U. (2004). Probing the unfolding region of ribonuclease a by site-directed mutagenesis. *European Journal of Biochemistry*, 271(20):4147–4156.

Korukottu, J., Schneider, R., Vijayan, V., Lange, A., Pongs, O., Becker, S., Baldus, M., and Zweckstetter, M. (2008). High-resolution 3d structure determination of kaliotoxin by solid-state nmr spectroscopy. *PLoS ONE*, 3(6):e2359.

Krebs, M. R. H., Bromley, E. H. C., and Donald, A. M. (2005). The binding of thioflavin-t to amyloid fibrils: localisation and implications. *J Struct Biol*, 149(1):30–37.

Krishnan, R. and Lindquist, S. L. (2005). Structural insights into a yeast prion illuminate nucleation and strain diversity. *Nature*, 435(7043):765–772.

Kuwajima, K. (1989). The molten globule state as a clue for understanding the folding and cooperativity of globular-protein structure. *Proteins*, 6(2):87–103.

Lange, A., Becker, S., Seidel, K., Giller, K., Pongs, O., and Baldus, M. (2005). A concept for rapid protein-structure determination by solid-state nmr spectroscopy. *Angew Chem Int Ed Engl*, 44(14):2089–92.

Lange, A., Gattin, Z., Van Melckebeke, H., Wasmer, C., Soragni, A., van Gunsteren, W. F., and Meier, B. H. (2009). A combined solid-state nmr and md characterization of the stability and dynamics of the het-s(218-289) prion in its amyloid conformation. *Chembiochem*, 10(10):1657–1665.

Lange, A., Giller, K., Hornig, S., Martin-Eauclaire, M.-F., Pongs, O., Becker, S., and Baldus, M. (2006). Toxin-induced conformational changes in a potassium channel revealed by solid-state nmr. *Nature*, 440(7086):959–962.

Lange, A., Luca, S., and Baldus, M. (2002). Structural constraints from proton-mediated rare-spin correlation spectroscopy in rotating solids. *J Am Chem Soc*, 124(33):9704–5.

Lange, A. and Meier, B. H. (2008). Fungal prion proteins studied by solid-state nmr. *C. R. Chimie*, 11(4-5):332–339.

Lange, A., Seidel, K., Verdier, L., Luca, S., and Baldus, M. (2003). Analysis of proton-proton transfer dynamics in rotating solids and their use for 3d structure determination. *J Am Chem Soc*, 125(41):12640–8.

Laskowski, R. A., Rullmannn, J. A., MacArthur, M. W., Kaptein, R., and Thornton, J. M. (1996). Aqua and procheck-nmr: programs for checking the quality of protein structures solved by nmr. *J Biomol NMR*, 8(4):477–486.

Laws, D. D., Bitter, H. M., Liu, K., Ball, H. L., Kaneko, K., Wille, H., Cohen, F. E., Prusiner, S. B., Pines, A., and Wemmer, D. E. (2001). Solid-state nmr studies of the secondary structure of a mutant prion protein fragment of 55 residues that induces neurodegeneration. *Proc Natl Acad Sci U S A*, 98(20):11686–11690.

Lazo, N. D. and Downing, D. T. (1998). Amyloid fibrils may be assembled from beta-helical protofibrils. *Biochemistry*, 37(7):1731–1735.

LeMaster, D. and Kushlan, D. (1996). Dynamical mapping of e. coli thioredoxin via 13c nmr relaxation analysis. *J Am Chem Soc*, 118(39):9255–9264.

LeMaster, D. M. and Richards, F. M. (1988). Nmr sequential assignment of escherichia coli thioredoxin utilizing random fractional deuteriation. *Biochemistry*, 27(1):142–50.

Lewandowski, J. R., De Paepe, G., and Griffin, R. G. (2007). Proton assisted insensitive nuclei cross polarization. *J Am Chem Soc*, 129(4):728–729.

Li, R. and Woodward, C. (1999). The hydrogen exchange core and protein folding. *Protein Sci*, 8(8):1571–90.

Linser, R., Fink, U., and Reif, B. (2008). Proton-detected scalar coupling based assignment strategies in mas solid-state nmr spectroscopy applied to perdeuterated proteins. *J Magn Reson*, 193(1):89–93.

Loquet, A., Bousset, L., Gardiennet, C., Sourigues, Y., Wasmer, C., Habenstein, B., Schütz, A., Meier, B. H., Melki, R., and Böckmann, A. (2009). Prion fibrils of ure2p assembled under physiological conditions contain highly ordered, natively folded modules. *J Mol Biol*, 394(1):108–118.

Lowe, I. (1959). Free induction decays of rotating solids. *Physical Review Letters*, 2(7):285–287.

Maddelein, M.-L. (2007). Infectious fold and amyloid propagation in podospora anserina. *Prion*, 1(1):44–47.

Maddelein, M.-L., Dos Reis, S., Duvezin-Caubet, S., Coulary-Salin, B., and Saupe, S. J. (2002). Amyloid aggregates of the het-s prion protein are infectious. *Proc Natl Acad Sci U S A*, 99(11):7402–7407.

Maji, S. K., Perrin, M. H., Sawaya, M. R., Jessberger, S., Vadodaria, K., Rissman, R. A., Singru, P. S., Nilsson, K. P. R., Simon, R., Schubert, D., Eisenberg, D., Rivier, J., Sawchenko, P., Vale, W., and Riek, R. (2009). Functional amyloids as natural storage of peptide hormones in pituitary secretory granules. *Science*, 325(5938):328–32.

Manolikas, T., Herrmann, T., and Meier, B. H. (2008). Protein structure determination from 13c spin-diffusion solid-state nmr spectroscopy. *J Am Chem Soc*, 130(12):3959–3966.

Mathiason, C. K., Powers, J. G., Dahmes, S. J., Osborn, D. A., Miller, K. V., Warren, R. J., Mason, G. L., Hays, S. A., Hayes-Klug, J., Seelig, D. M., Wild, M. A., Wolfe, L. L., Spraker, T. R., Miller, M. W., Sigurdson, C. J., Telling, G. C., and Hoover, E. A. (2006). Infectious prions in the saliva and blood of deer with chronic wasting disease. *Science*, 314(5796):133–136.

Morell, M., Bravo, R., Espargaro, A., Sisquella, X., Aviles, F. X., Fernandez-Busquets, X., and Ventura, S. (2008). Inclusion bodies: specificity in their aggregation process and amyloid-like structure. *Biochim Biophys Acta*, 1783(10):1815–1825.

Moreno-Hagelsieb, G. and Latimer, K. (2008). Choosing blast options for better detection of orthologs as reciprocal best hits. *Bioinformatics*, 24(3):319–24.

Morris, G. A. and Freeman, R. (1979). Enhancement of nuclear magnetic resonance signals by polarization transfer. *J Am Chem Soc*, 101(3):760–762.

Nelson, R. and Eisenberg, D. (2006). Structural models of amyloid-like fibrils. *Adv Protein Chem*, 73:235–282.

Ohgushi, M. and Wada, A. (1983). 'molten-globule state': a compact form of globular proteins with mobile side-chains. *FEBS Lett*, 164(1):21–24.

Pace, C., Hebert, E., Shaw, K., Schell, D., Both, V., Krajcikova, D., Sevcik, J., Wilson, K., Dauter, Z., Hartley, R., and Grimsley, G. (1998). Conformational stability and thermodynamics of folding of ribonucleases sa, sa2 and sa3. *Journal of Molecular Biology*, 279(1):271–286.

Paravastu, A. K., Petkova, A. T., and Tycko, R. (2006). Polymorphic fibril formation by residues 10-40 of the alzheimer's beta-amyloid peptide. *Biophys J*, 90(12):4618–4629.

Parry, D., Jenkinson, P., and McLeod, L. (1995). Fusarium ear blight (scab) in small grain cereals - a review. *Plant Pathology*, 44(2):207–238.

Pervushin, K., Riek, R., Wider, G., and Wüthrich, K. (1997). Attenuated t2 relaxation by mutual cancellation of dipole-dipole coupling and chemical shift anisotropy indicates an avenue to nmr structures of very large biological macromolecules in solution. *Proc Natl Acad Sci U S A*, 94(23):12366–71.

Petkova, A. T., Ishii, Y., Balbach, J. J., Antzutkin, O. N., Leapman, R. D., Delaglio, F., and Tycko, R. (2002). A structural model for alzheimer's beta-amyloid fibrils based on experimental constraints from solid state nmr. *Proc Natl Acad Sci U S A*, 99(26):16742–16747.

Petkova, A. T., Leapman, R. D., Guo, Z., Yau, W.-M., Mattson, M. P., and Tycko, R. (2005). Self-propagating, molecular-level polymorphism in alzheimer's beta-amyloid fibrils. *Science*, 307(5707):262–265.

Petkova, A. T., Yau, W.-M., and Tycko, R. (2006). Experimental constraints on quaternary structure in alzheimer's beta-amyloid fibrils. *Biochemistry*, 45(2):498–512.

Pines, A., Gibby, M., and Waugh, J. (1973). Proton-enhanced nmr of dilute spins in solids. *Journal of Chemical Physics*, 59(2):569–590.

Prusiner, S. B. (1982). Novel proteinaceous infectious particles cause scrapie. *Science*, 216(4542):136–144.

Prusiner, S. B., Scott, M. R., DeArmond, S. J., and Cohen, F. E. (1998). Prion protein biology. *Cell*, 93(3):337–348.

Riek, R., Hornemann, S., Wider, G., Billeter, M., Glockshuber, R., and Wüthrich, K. (1996). Nmr structure of the mouse prion protein domain prp(121-321). *Nature*, 382(6587):180–2.

Rienstra, C., Hohwy, M., Hong, M., and Griffin, R. (2000). 2d and 3d n-15-c-13-c-13 nmr chemical shift correlation spectroscopy of solids: Assignment of mas spectra of peptides. *Journal of the American Chemical Society*, 122(44):10979–10990.

Ritter, C., Maddelein, M.-L., Siemer, A. B., Luhrs, T., Ernst, M., Meier, B. H., Saupe, S. J., and Riek, R. (2005). Correlation of structural elements and infectivity of the het-s prion. *Nature*, 435(7043):844–848.

Rizet, G. (1952). Les phénomènes de barrage chez Podospora anserina: Analyse genetique des barrages entre les souches s et S. *Rev. Cytol. Biol. Veg*, 13:51–92.

Sabate, R., Baxa, U., Benkemoun, L., Sanchez de Groot, N., Coulary-Salin, B., Maddelein, M., Malato, L., Ventura, S., Steven, A., and Saupe, S. (2007). Prion and non-prion amyloids of the het-s prion forming domain. *J Mol Biol*, 370(4):768–783.

Santoro, M. and Bolen, D. (1988). Unfolding free energy changes determined by the linear extrapolation method. 1. unfolding of phenylmethanesulfonyl .alpha.-chymotrypsin using different denaturants. *Biochemistry*, 27(21):8063–8068.

Saupe, S. J. (2000). Molecular genetics of heterokaryon incompatibility in filamentous ascomycetes. *Microbiol Mol Biol Rev*, 64(3):489–502.

Saupe, S. J. (2007). A short history of small s: a prion of the fungus podospora anserina. *Prion*, 1(2):110–115.

Sawaya, M. R., Sambashivan, S., Nelson, R., Ivanova, M. I., Sievers, S. A., Apostol, M. I., Thompson, M. J., Balbirnie, M., Wiltzius, J. J. W., McFarlane, H. T., Madsen, A. O., Riekel, C., and Eisenberg, D. (2007). Atomic structures of amyloid cross-beta spines reveal varied steric zippers. *Nature*, 447(7143):453–457.

Scholz, I., Huber, M., Manolikas, T., Meier, B. H., and Ernst, M. (2008). Mirror recoupling and its application to spin diffusion under fast magic-angle spinning. *Chem Phys Lett*, 460(1-3):278–283.

Scholz, I., Meier, B. H., and Ernst, M. (2007). Operator-based triple-mode floquet theory in solid-state nmr. *J Chem Phys*, 127(20):204504.

Scholz, I., Meier, B. H., and Ernst, M. (2009). Mirror-cp: A proton-only experiment for the measurement of c-13 spin diffusion. *Chemical Physics Letters*, 479(4-6):296–299.

Sen, A., Baxa, U., Simon, M. N., Wall, J. S., Sabate, R., Saupe, S. J., and Steven, A. C. (2007). Mass analysis by scanning transmission electron microscopy and electron diffraction validate predictions of stacked beta-solenoid model of het-s prion fibrils. *J Biol Chem*, 282(8):5545–5550.

Shewmaker, F., Kryndushkin, D., Chen, B., Tycko, R., and Wickner, R. B. (2009). Two prion variants of sup35p have in-register parallel beta-sheet structures, independent of hydration. *Biochemistry*, 48(23):5074–82.

Shewmaker, F., Wickner, R. B., and Tycko, R. (2006). Amyloid of the prion domain of sup35p has an in-register parallel beta-sheet structure. *Proc Natl Acad Sci U S A*, 103(52):19754–19759.

Siemer, A. B. (2006). *Structure Determination With Solid-State NMR: The HET-s Prion Protein*. PhD thesis, ETH Zurich.

Siemer, A. B., Arnold, A. A., Ritter, C., Westfeld, T., Ernst, M., Riek, R., and Meier, B. H. (2006a). Observation of highly flexible residues in amyloid fibrils of the het-s prion. *J Am Chem Soc*, 128(40):13224–13228.

Siemer, A. B., Ritter, C., Ernst, M., Riek, R., and Meier, B. H. (2005). High-resolution solid-state nmr spectroscopy of the prion protein het-s in its amyloid conformation. *Angew Chem Int Ed Engl*, 44(16):2441–2444.

Siemer, A. B., Ritter, C., Steinmetz, M. O., Ernst, M., Riek, R., and Meier, B. H. (2006b). 13c, 15n resonance assignment of parts of the het-s prion protein in its amyloid form. *J Biomol NMR*, 34(2):75–87.

Spera, S. and Bax, A. (1991). Empirical correlation between protein backbone conformation and c.alpha. and c.beta. 13c nuclear magnetic resonance chemical shifts. *J Am Chem Soc*, 113(14):5490–5492.

Steinbacher, S., Baxa, U., Miller, S., Weintraub, A., Seckler, R., and Huber, R. (1996). Crystal structure of phage p22 tailspike protein complexed with salmonella sp. o-antigen receptors. *Proc Natl Acad Sci U S A*, 93(20):10584–10588.

Straus, S. K., Bremi, T., and Ernst, R. R. (1998). Experiments and strategies for the assignment of fully 13c/15n-labelled polypeptides by solid state nmr. *J Biomol NMR*, 12(1):39–50.

Szeverenyi, N. M., Sullivan, M. J., and Maciel, G. E. (1982). Observation of spin exchange by two-dimensional fourier transform 13c cross polarization-magic-angle spinning. *J Magn Reson (1969)*, 47(3):462–475.

Takegoshi, K., Nakamura, S., and Terao, T. (2001). C-13-h-1 dipolar-assisted rotational resonance in magic-angle spinning nmr. *Chem Phys Lett*, 344:631–637.

Toyama, B. H., Kelly, M. J. S., Gross, J. D., and Weissman, J. S. (2007). The structural basis of yeast prion strain variants. *Nature*, 449(7159):233–237.

Turcq, B., Deleu, C., Denayrolles, M., and Begueret, J. (1991). Two allelic genes responsible for vegetative incompatibility in the fungus podospora anserina are not essential for cell viability. *Mol Gen Genet*, 228(1-2):265–269.

Tycko, R. (2006). Molecular structure of amyloid fibrils: insights from solid-state nmr. *Q Rev Biophys*, 39(1):1–55.

Tycko, R., Sciarretta, K. L., Orgel, J. P. R. O., and Meredith, S. C. (2009). Evidence for novel beta-sheet structures in iowa mutant beta-amyloid fibrils. *Biochemistry*, 48(26):6072–84.

Ulrich, E. L., Akutsu, H., Doreleijers, J. F., Harano, Y., Ioannidis, Y. E., Lin, J., Livny, M., Mading, S., Maziuk, D., Miller, Z., Nakatani, E., Schulte, C. F., Tolmie, D. E., Kent Wenger, R., Yao, H., and Markley, J. L. (2008). Biomagresbank. *Nucleic Acids Res*, 36(Database issue):D402–408.

Ulrich, E. L., Markley, J. L., and Kyogoku, Y. (1989). Creation of a nuclear magnetic resonance data repository and literature database. *Protein Seq Data Anal*, 2(1):23–37.

Van Melckebeke, H., Schanda, P., Gath, J., Wasmer, C., Verel, R., Lange, A., Meier, B. H., and Böckmann, A. (2011). Probing water accessibility in het-s(218-289) amyloid fibrils by solid-state nmr. *J Mol Biol*, 405(3):765–72.

Van Melckebeke, H., Wasmer, C., Lange, A., Ab, E., Loquet, A., Böckmann, A., and Meier, B. H. (2010). Atomic-resolution three-dimensional structure of het-s(218-289) amyloid fibrils by solid-state nmr spectroscopy. *J Am Chem Soc*, 132(39):13765–75.

Ventura, S. and Villaverde, A. (2006). Protein quality in bacterial inclusion bodies. *Trends Biotechnol*, 24(4):179–185.

Verel, R., Ernst, M., and Meier, B. H. (2001). Adiabatic dipolar recoupling in solid-state nmr: the dream scheme. *J Magn Reson*, 150(1):81–99.

Vilar, M., Chou, H.-T., Luhrs, T., Maji, S. K., Riek-Loher, D., Verel, R., Manning, G., Stahlberg, H., and Riek, R. (2008). The fold of alpha-synuclein fibrils. *Proc Natl Acad Sci U S A*, 105(25):8637–8642.

Wadsworth, J. D. F., Joiner, S., Linehan, J. M., Asante, E. A., Brandner, S., and Collinge, J. (2008). The origin of the prion agent of kuru: molecular and biological strain typing. *Philos Trans R Soc Lond B Biol Sci*, 363(1510):3747–3753.

Wang, L., Maji, S. K., Sawaya, M. R., Eisenberg, D., and Riek, R. (2008). Bacterial inclusion bodies contain amyloid-like structure. *PLoS Biol*, 6(8):e195.

Wang, Y. and Jardetzky, O. (2002). Probability-based protein secondary structure identification using combined nmr chemical-shift data. *Protein Sci*, 11(4):852–861.

Wasmer, C., Benkemoun, L., Sabaté, R., Steinmetz, M. O., Coulary-Salin, B., Wang, L., Riek, R., Saupe, S. J., and Meier, B. H. (2009a). Solid-state nmr spectroscopy reveals that e. coli inclusion bodies of het-s(218-289) are amyloids. *Angew Chem Int Ed Engl*, 48(26):4858–4860.

Wasmer, C., Guntert, P., Saupe, S. J., and Meier, B. H. (2011). The structure of fghet-s(218-289). *in preparation*, -(-):–.

Wasmer, C., Lange, A., Van Melckebeke, H., Siemer, A. B., Riek, R., and Meier, B. H. (2008a). Amyloid fibrils of the het-s(218-289) prion form a beta solenoid with a triangular hydrophobic core. *Science*, 319(5869):1523–1526.

Wasmer, C., Schütz, A., Loquet, A., Buhtz, C., Greenwald, J., Riek, R., Böckmann, A., and Meier, B. H. (2009b). The molecular organization of the fungal prion het-s in its amyloid form. *J Mol Biol*, 394(1):119–127.

Wasmer, C., Soragni, A., Sabate, R., Lange, A., Riek, R., and Meier, B. H. (2008b). Infectious and noninfectious amyloids of the het-s(218-289) prion have different nmr spectra. *Angew Chem Int Ed Engl*, 47(31):5839–5841.

Wasmer, C., Zimmer, A., Sabaté, R., Soragni, A., Saupe, S. J., Ritter, C., and Meier, B. H. (2010). Structural similarity between the prion domain of het-s and a homologue can explain amyloid cross-seeding in spite of limited sequence identity. *J Mol Biol*, 402(2):311–25.

Wells, G. A., Scott, A. C., Johnson, C. T., Gunning, R. F., Hancock, R. D., Jeffrey, M., Dawson, M., and Bradley, R. (1987). A novel progressive spongiform encephalopathy in cattle. *Vet Rec*, 121(18):419–420.

Wickner, R., Shewmaker, F., Kryndushkin, D., and Edskes, H. (2008). Protein inheritance (prions) based on parallel in-register beta-sheet amyloid structures. *Bioessays*, 30(10):955–964.

Wickner, R. B. (1994). [ure3] as an altered ure2 protein: evidence for a prion analog in saccharomyces cerevisiae. *Science*, 264(5158):566–569.

Wickner, R. B., Edskes, H. K., Shewmaker, F., and Nakayashiki, T. (2007). Prions of fungi: inherited structures and biological roles. *Nat Rev Microbiol*, 5(8):611–618.

Will, R. G., Ironside, J. W., Zeidler, M., Cousens, S. N., Estibeiro, K., Alperovitch, A., Poser, S., Pocchiari, M., Hofman, A., and Smith, P. G. (1996). A new variant of creutzfeldt-jakob disease in the uk. *Lancet*, 347(9006):921–925.

Wishart, D. S., Bigam, C. G., Holm, A., Hodges, R. S., and Sykes, B. D. (1995). 1h, 13c and 15n random coil nmr chemical shifts of the common amino acids. i. investigations of nearest-neighbor effects. *J Biomol NMR*, 5(1):67–81.

Wishart, D. S. and Nip, A. M. (1998). Protein chemical shift analysis: a practical guide. *Biochem Cell Biol*, 76(2-3):153–163.

Wittekind, M. and Mueller, L. (1993). Hncacb, a high-sensitivity 3d nmr experiment to correlate amide-proton and nitrogen resonances with the alpha- and beta-carbon resonances in proteins. *J Magn Reson, Ser B*, 101(2):201–205. Journal of Magnetic Resonance Series B.

Yoder, M. D., Keen, N. T., and Jurnak, F. (1993). New domain motif: the structure of pectate lyase c, a secreted plant virulence factor. *Science*, 260(5113):1503–1507.

Zahn, R., Liu, A., Lührs, T., Riek, R., von Schroetter, C., López García, F., Billeter, M., Calzolai, L., Wider, G., and Wüthrich, K. (2000). Nmr solution structure of the human prion protein. *Proc Natl Acad Sci U S A*, 97(1):145–50.

Zhou, D. H., Shah, G., Cormos, M., Mullen, C., Sandoz, D., and Rienstra, C. M. (2007). Proton-detected solid-state nmr spectroscopy of fully protonated proteins at 40 khz magic-angle spinning. *J Am Chem Soc*, 129(38):11791–11801.

Acknowledgments

As most advances in contemporary science, the accomplishments presented in this thesis were only possible with the invaluable contributions and the support of a number of collaborating scientists.

First, I want to thank my supervisor, **Prof. Beat Meier** for his vast support, both financially and—more importantly—scientifically in numberless fruitful discussions and for giving me the opportunity to work in his research group.

I thank **Prof. Roland Riek** not only for co-examining this thesis, but also for discussions and collaborations on several highly interesting research projects.

Prof. Sebastian Hiller and **Ansgar Siemer**, who invested a lot of their time to introduce me to biological solution NMR and solid-state NMR, respectively.

A very important part of this work, the determination of the HET-s(218-289) structure, was a team-effort with two great researchers, **Hélène Van Melckebeke** and **Adam Lange**, who contributed a vast amount of knowledge and time.

Almost all other projects described in this thesis also involved collaborations (as described in detail in the beginning of the according chapters). For this I am thankful to **Agnes Zimmer, Prof. Christiane Ritter; Raimon Sabaté, Laura Benkemoun, Bénédicte Coulary-Salin, Prof. Sven Saupe; Alice Soragni, Caroline Buhtz, Jason Greenwald, Lei Wang; Anne Schütz; Prof. Peter Güntert; Birgit Habenstein, Antoine Loquet, Anja Böckmann;** and **Michel Steinmetz**.

I also want to thank the whole **solid-state NMR group** at the ETH Zurich with all its current and former members for the countless highly scientific and less so discussions in the coffee room and for having a good time with you in Zurich.

Finally, I thank all of my friends from within and out of Zurich, in particular the last survivors of the 'Freitagsrunde'. I want to especially thank my brothers and parents for their love and constant support. Without all your encouragement and aid this work would not have been possible.

Acknowledgments

Die VDM Verlagsservicegesellschaft sucht für wissenschaftliche Verlage abgeschlossene und herausragende

Dissertationen, Habilitationen, Diplomarbeiten, Master Theses, Magisterarbeiten usw.

für die kostenlose Publikation als Fachbuch.

Sie verfügen über eine Arbeit, die hohen inhaltlichen und formalen Ansprüchen genügt, und haben Interesse an einer honorarvergüteten Publikation?

Dann senden Sie bitte erste Informationen über sich und Ihre Arbeit per Email an *info@vdm-vsg.de*.

Sie erhalten kurzfristig unser Feedback!

VDM Verlagsservicegesellschaft mbH
Dudweiler Landstr. 99
D - 66123 Saarbrücken

Telefon +49 681 3720 174
Fax +49 681 3720 1749

www.vdm-vsg.de

Die VDM Verlagsservicegesellschaft mbH vertritt

Printed by Books on Demand GmbH, Norderstedt / Germany